Environmental science laboratory manual

Environmental
science laboratory manual

MAURICE A. STROBBE, Ph.D.

Professor and Chairman, Department of Biology, Iowa Wesleyan College,
Mt. Pleasant, Iowa; Formerly Head of Department of Biology, Black Hawk College,
Moline, Illinois; Consultant for the Educational Affairs Division, Argonne National
Laboratories, Argonne, Illinois

THE C. V. MOSBY COMPANY
Saint Louis   1972

Preface

The objective of this manual is to provide a set of basic analytical procedures commonly used to determine the quality of our environment. These procedures are designed to be used for an introductory course in environmental science. The exercises are based on quantitative and qualitative determinations, with procedures explicit enough to allow them to be performed by the nonscience or beginning science student. The diversity of experiments is such that they will serve as laboratory exercises for any course dealing with the subject of environmental quality, regardless of the texts or references used, in biology, chemistry, or general science courses, as well as in new environmental science courses.

Most of the material used for the exercises will be collected from the immediate locale. This will provide the student with an awareness of the pollution parameters within his own community, their sources, and possible solutions. The necessary apparatus is common to most science departments, thus requiring minimal expenditure.

This manual is divided into six parts: Part I, Particulate Matter and Chemical Parameters Affecting Air Quality; Part II, Chemical Analysis of Water; Part III, Microbiological Examination of Water; Part IV, Physical Parameters of Water; Part V, Identification of Pesticide Residues by Thin-Layer Chromatography; Part VI, Appendixes: Glossary, Weight and Volume Equivalents, Effects of Particulate Matter, Effects of Sulfur Dioxide, Effects of Nitrogen Dioxide, Effects of Carbon Monoxide, Air Quality Standards (by the Environmental Protection Agency), Surface Water Criteria for Public Water Supplies, Air Pollution Series of Technical Publications, and Federal Pollution Control Agencies.

A discussion of the parameters is included at the beginning of most of the exercises or at the beginning of major sections. These discussions point out the significance of the parameters to our environmental health. They also serve to support the instructor and student in discussions of results. At the end of each exercise there is a Result Table to be completed by the student. The table will supply criteria supportive to the application of deductive and inductive reasoning to the questions that follow each experiment.

More exercises are included than can be completed in a one-semester course; therefore the instructor can select the exercises most pertinent to his locale.

The exercises provided in each part have been adopted and modified from Standard Methods of Examination of Water and Wastewater (editions 12 and 13, New York, 1965 and 1971, American Public Health Association); Water Analysis Procedures have been adapted and modified from Hach Chemical Company; Air Analysis Procedures have been recommended by Gelman Instrument Company, Millipore Corporation, National Environmental Instruments, Inc., and other state and industrial analytical sources as indicated. Many of the exercises involve prepackaged materials from these suppliers, thus preventing the course from becoming a chemistry course for instructors and students alike and keeping the preclass preparation to a minimum.

The exercises were performed by environmental classes and National Science Foundation participants at the University of Iowa, Iowa City, prior to the publication of this manual.

I am grateful for the support and cooperation received from Mr. Jerry Baird, Dr. Bruce Murray, Mr. Arthur Ferreri, and Dr. Rosalind Klaas, a consultant of Argonne National Laboratories. Many of the experiments of this manual are an outgrowth from attending and working at the various workshops at the Argonne Center for Educational Affairs, Argonne Laboratories (AEC), Argonne, Illinois.

Maurice A. Strobbe, Ph.D.

Contents

Environmental science laboratory manual

PART I

Particulate matter and chemical parameters affecting air quality

Air is normally made up of a combination of gases and small concentrations of naturally occuring dust particles, plant pollen, spores, and necessary microorganisms. Therefore, air pollution is the increasing of the quantity of normal constituents and the addition of others, which may be toxic, as the result of man's activities. Many toxic compounds such as pesticides, herbicides, fungicides, bactericides, and heavy-metal vapors can be found in association with both suspended and settleable solids. Thus particulate matter increases the exposure of plants, animals, and man to the toxic effects of these compounds.

Air and its contents are not normally a propagating medium for bacteria and viruses. However, they serve as an excellent transport medium for pathogenic and nonpathogenic organisms found in air. Many respiratory diseases have been shown to be transmitted by air and the particulate matter within the air, such as rheumatic fever, tuberculosis, diphtheria, and scarlet fever. When the air is polluted with certain chemical compounds or gases, they may react to cause damage to vegetation; shorten the life of clothing; damage wood and metal; cause skin irritation, coughing, burning of eyes and throat, death to animals; and increase the death rate among people who have minor respiratory ailments.

Lead from gasoline is released into the air from automobile exhausts. It is a cumulative poison; that is, it is not eliminated as fast as it is taken into the body. It is readily absorbed by food, water, or air. There is evidence that lead has injurious effects on the blood, kidneys, and liver of man.

1
Particulate matter and lead in air

Particulate matter may be classified as suspended and settleable. The suspended particulate is small, in the range of 1 to 100 microns in size. The surface area of a small particle is greater in proportion to volume than that of a large-sized particle. The small particle will stay suspended in air for long periods of time, whereas the large settleable particulate will settle close to its source. The quantity of particulate matter in air is determined by weight units of grams, milligrams, micrograms, and the number of particles present per unit of volume of air. In concentrations of over 200 $\mu g/m^3$, there is a decrease in visibility. Photochemical reactions between particulate matter and gases produces compounds that contribute to the increase in mortality of people with chronic respiratory diseases. This quality also contributes to the evidence of total morbidity and cardiovascular diseases among middle-class individuals over 50 years of age.

The composition of the particulate is directly related to the industry and population of a geographic location.

The number of particles per volume of air is determined in this exercise. However, if weights are wanted, weigh the filter pad both prior to and after collecting the sample. Make the appropriate calculations and record them in the result table.

Materials

1. Millipore vacuum-pump assembly (Fig. 1) (Millipore Corporation, Bedford, Mass.)
2. Type AA (0.8-micron pore size) Millipore filter
3. Vacuum source or Millipore vacuum-pressure pump
4. Petri dishes, eye droppers, tweezers
5. Tetrahydroxyquinone

Procedural information

The volume of air drawn through the filter is approximately 1 liter per minute (Lpm) per inch of vacuum. The following scale is recommended for determining the sampling area:

Room air—10 cubic feet = 10 Lpm for 28 minutes.

Country air—5 cubic feet = 10 Lpm for 14 minutes.

City air—1 cubic foot = 1 Lpm for 28 minutes.

Factory air—0.5 cubic foot = Lpm for 28 minutes.

There are about 28.31 L per cubic foot of air. It is possible to increase the vacuum, thus reducing the sampling time.

Procedure

1. Select four sampling areas, such as a classroom, a parking lot, a classroom window, or a garage.
2. Set up the filtering system as indicated in Fig. 1, using the proper filter.
3. Determine the sampling time for the area selected.
4. *If* the number of cubic feet is to be determined, adjust the vacuum pressure to the recommendations. Post the cubic feet in the result table.
5. *If* air volume is wanted, adjust the vacuum and record the time. Post the volume in the result table.
6. After collecting the air sample, remove the filter with a pair of tweezers and put it in a clean petri dish and cover.
7. Using a microscope set at low power, a dissecting microscope, or a hand lens, count the visible particles.
8. It is not necessary to count all the particles observed on the filter. Count 10 to 20 grid squares and use the following

Fig. 1. Millipore vacuum-pressure pump XX60 000 00 and assembly.

formula to determine the number of particles.

$$\frac{145}{\text{Number of squares counted}} \times \text{Count} = \text{Total count}$$

9. Post your count and the counts for the other samples in the result table.
10. Refer to the Introduction and Appendix, pp. 127 and 131, for added information supportive to this exercise.

Lead determination procedure

1. Remove the filter from the assembly.
2. Add a few drops of tetrahydroxyquinone to precipitate lead.
3. A red color will appear if lead particles are present. Estimate the concentration by the degree of redness, that is, dark, medium, or light.
4. Post the estimates in the result table.

Result table

Sample areas	Cubic feet of air	Liters of air	Total count	Lead present	Remarks
1					
2					
3					
4					

Questions

Compare your total particulate count with state or federal standards. Post the differences in the remark section of the table. Do the same for the lead.

Determine the possible sources for the particulate matter and lead for each sample area.

Are the quantities of particulate matter and lead present likely to have a physiological effect on the environment? Explain the effects on man, plants, and animals.

How can the quantity of particulate matter and lead be controlled in the sampled environment?

2
Microorganisms of air and particulate matter

Microorganisms may adhere to and be carried by many kinds of particulate matter particles. When these organisms are introduced into the air, they may be transported many miles by air currents. The distribution of airborne disease organisms is determined by the atmospheric conditions, quantity, size, and nature of particulate matter, which directly affects the number of people infected with respiratory diseases. Therefore, the spread of many respiratory diseases such as scarlet fever, diphtheria, mumps, and influenza viruses can be attributed to the contamination or pollution of air. The presence and number of airborne bacteria and fungal spores in the air can be determined by the following experiments:

TEST 1—THE PRESENCE OF MICROORGANISMS
Materials

1. Petri dishes with nutrient agar or tryptone lactose yeast agar (see the *Difco*

Manual or any microbiology laboratory manual for preparation).

Procedure

1. Select four locations to expose the petri dishes.
2. Identify the locations. Mark each petri dish.

1 _____.

2 _____.

3 _____.

4 _____.

3. Expose the petri dishes for 15 minutes in each location and replace the lid.
4. Invert the petri dishes and incubate at 37° C for 24 to 48 hours.
5. Record the number of bacteria, number of fungi, and total number of colonies counted in the result table.

Result table

Locality	Number of colonies	Number of bacteria colonies	Number of fungus colonies	Remarks
1				
2				
3				
4				

Questions

Are there more bacteria than fungus colonies?

Will all the different kinds of organisms be represented on your culture? Explain.

Can this be counted as a pollution-parameter test? Explain.

Explain the possible presence of pathogens.

TEST 2—AIRBORNE ORGANISMS PER VOLUME OF AIR

INSTRUCTOR: This experiment utilizes the procedures developed by the Millipore Corporation, Bedford, Mass. It is recommended that the students work in groups. Each group can sample a different locality. The following procedures are based upon three sampling areas. (Instructions for the preparation of the media may be found in the *Difco Manual* or any microbiology laboratory manual.)

Materials

3 Special All-Glass impingers (12.5 Lpm)
1 Vacuum-pressure pump
3 Syringes and two-way valve
3 Stainless steel forceps
3 Clinical monitors
Sterile sampling tubes
Hose clamp
Ring stand and clamp
Gelatin
Brain heart infusion broth (dehydrated)
Disodium phosphate (anhydrous)
Antifoam AF emulsion (Dow-Corning)
Type AA (0.8 micron) Millipore filter pads (optional)
Impingement fluid: Dissolve the following in 1 liter of water: 2 grams of gelatin, 4 grams of disodium phosphate (anhydrous), 37 grams of brain heart infusion, and 0.1 ml of Antifoam AF emulsion. Autoclave for 15 minutes at 15 pounds per square inch (psi).

Procedure

1. Sterilize the All-Glass impinger and clamp it to a ring stand, as shown in Fig. 2, *A*.

2. Remove the top (blue) plug from a clinical monitor and insert the nylon adapter end of a sterile sampling tube. Slip the other end of the sampling tube over the tubulation on the bottom of the impinger and clamp off the tube.

3. Pour 30 ml of sterile impingement fluid into the impinger.

4. Attach the impinger to a vacuum source (minimum 18 inches of Hg required) and turn on the vacuum. Approximately 12.5 Lpm will flow through the apparatus, yielding a sample volume of 10 cubic feet in 23 minutes.

5. After the appropriate time period, turn off the vacuum. Remove the bottom (red) plug from the monitor and attach the syringe and valve (Fig. 2, *C*). The syringe and valve need not be sterile.

6. Release the hose clamp and draw out all but 1 or 2 ml of the impingement fluid from the impinger.

7. Pour 10 ml of sterile broth medium into the impinger and draw it all through the monitor. *Stop* filtering the *instant* the last few drops of medium disappear from the filter surface.

8. Detach the monitor and replace the plugs.

9. Place the monitor in an inverted position in an incubator at 37° C for 24 to 48 hours.

10. After the incubation period, count the colonies (bacteria and fungi) growing on the monitor grid pad. (Do not open the monitor to count the colonies.)

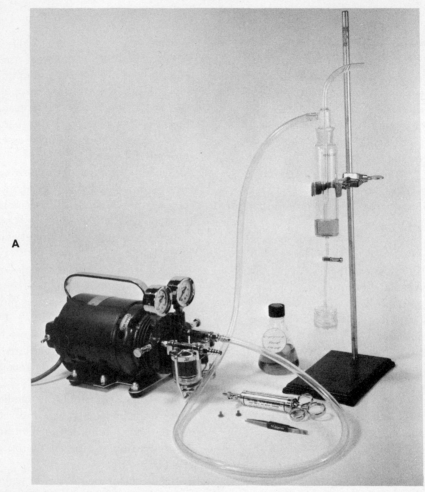

A

Fig. 2. A, Impinger, with 30 ml of impingement fluid, is set up in sampling area with clamped-off monitor on sampling tube. **B,** Viable recovery approaches 100% as organisms are impinged into fluid at near sonic velocity. **C,** Impinger fluid is drawn through the monitor to concentrate the organisms.

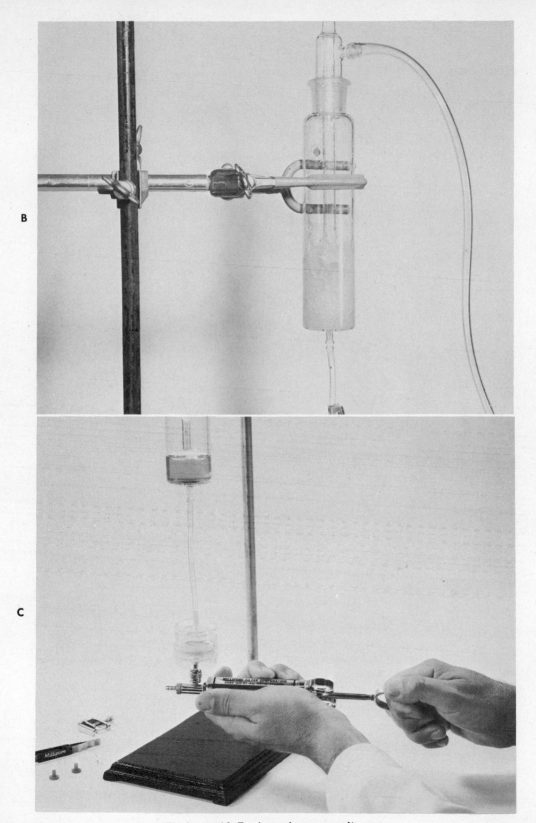

B

C

Fig. 2, cont'd. For legend see opposite page.

Result table

Population counts represent the number of microorganisms present. When approximately 12.5 Lpm of air flows through the device, a sample volume of 10 cubic feet is yielded in 23 minutes.

Sample	Number of bacteria	Number of fungi	Number of other	Different color bacteria	Different color fungi	Total
1						
2						
3						
4						

Questions

Itemize the different colors of bacteria colonies.

Itemize the different colors of fungi colonies.

What pathogenic effects are some of these organisms likely to have on man?

What are the possible sources of the microorganisms found in the air sample?

Considering the areas from which the samples were collected, account for the differences in the microorganism population.

Should the microorganisms of air be considered an environmental quality parameter? Explain.

3
Carbon monoxide determination

Carbon monoxide (CO) is a by-product of the combustion of gasoline and diesel oils. It is toxic to man and animals. It reacts with compounds and gases in air to form additional toxic compounds. As little as 5 parts per million (ppm) effects reflex changes in the higher nerve centers. In concentrations of 30 ppm, visual and mental acuities are affected. Levels as high as 100 to 200 ppm have been measured in many cities. At these levels, many older people with minor ailments are affected with headaches, nausea, and dimness of vision.

Carbon monoxide is also a major contributor to smog, thereby affecting the health of people, plants, animals, and materials. See Appendix, pp. 130 and 131.

Materials

The equipment and materials used for this test can be supplied by National Environmental Instruments, Inc., Fall River, Mass.

1. Unico Precision Gas Detector Model 400 (one detector may be used by several students)
2. Carbon Monoxide Color Intensity Detector tubes, No. 106c
3. Carbon Monoxide Color standards (Unico)

Procedural considerations

CO forms an additive compound with palladium sulfate to form ammonium molybdate, which yields molybdenum blue. The color standard chart and waiting-time calibration are based on a tube temperature of 15° C (59° F), not gat temperature. If the temperature is higher or lower than 15° C, *waiting time* should be adjusted and the reading

of the color standard chart must be corrected according to the temperature correction table at the bottom of the color standard chart.

A prolonged *sampling* time provides a low concentration reading; a shortened sampling time usually gives a high concentration reading. If a sampling time of 2 minutes is used, then 25, 50, 75, 150, and 250 ppm readings are obtainable. If the sampling time of 5 seconds is used, 1200, 1800, 3600, and 6000 ppm readings are obtainable.

Procedure

1. Select four sampling areas, such as a classroom, a parking lot, the immediate area of an auto tail pipe, and a garage.
2. To sample, break the tips off a detector tube, bending each tube end in the tube-tip breaker of the pump (Fig. 3).
3. Insert the tube tip marked with the red dot securely into the pump inlet.
4. Make certain the pump handle is all the way in. Align the guide marks on the shaft and the back plate of the pump.
5. Pull the handle all the way out. Lock it with a half turn. Wait exactly 30 seconds. The sampling time should be counted from the precise time the handle is pulled out.
6. Remove the detector tube from the pump inlet. For tube temperatures of 59° and 68° F, wait 2 minutes for the discoloration to develop. For other temperatures, consult the temperature correction table. The *waiting time* must be counted from the precise moment the sampling is completed.
7. After a 2-minute *waiting period,* compare the discoloration of the reagent nearest to the red dot with the color standard chart to get the concentration in ppm.

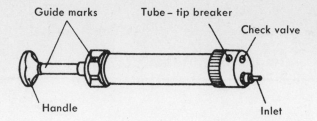

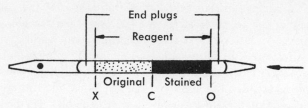

Fig. 3. Parts identification of the Unico pump and the gas detector tube.

8. At concentrations below 100 ppm, up to, or above 1000 ppm, use a longer or shorter sampling time. The true concentration can be determined from the following formula:

9. Post the ppm determinations in the result table.

10. Refer to the Introduction and Appendix, pp. 130 and 131, for information on carbon monoxide.

$$\text{True concentration} = \text{Reading from the color standard} \times \frac{30 \text{ seconds}}{\text{Other sampling time}}$$

Result table

Sample	Concentration ppm	TLV ppm	Difference	Remarks
1		50		
2		50		
3		50		
4		50		

NOTE: TLV = threshold limit value of 50 ppm for a daily 8-hour exposure at 25° C and 760 mm Hg for 1968.

Questions

What are some of the physiological effects of high concentrations of carbon monoxide?

How does carbon monoxide interfere with the body's normal processes?

What are the major sources of carbon monoxide? How can these be controlled?

Is it safe to work or live in the areas where the samples were taken? Why or why not?

Do high levels of carbon monoxide have an effect on organisms other than humans? Explain.

4
Carbon dioxide determination

There is little evidence today that the increasing level of carbon dioxide is causing adverse conditions in our environment. The major sources of CO_2 in our atmosphere are from burning fossil fuel and animal respiration. There has been an increase of about 14% in carbon dioxide within the past 50 years. The major concern has been the possible effect of a continual increase of carbon dioxide on global temperature. It has been shown that as carbon dioxide increases, the heat loss into the upper atmosphere through radiation from the surface decreases. This is known as the "greenhouse" effect. However, there have been no substantial changes in the earth's temperature to support this theory. In fact, during the recent period of rapid increase in CO_2, the earth has become somewhat cooler. The heating effects of CO_2 must therefore, be counteracted by man's activities and the increase of CO_2 absorption by the ocean. There is also concern over the possible photochemical activity of large concentrations of carbon dioxide. Carbon dioxide reacts with water vapor molecules to form carbonic acid, which may cause deterioration of stone buildings and structures. It also serves as an indicator of the burning (or consumption) of fossil fuel.

Material

The equipment and materials for this test can be supplied by National Environmental Instruments, Inc., Fall River, Mass.
1. Unico Precision Gas Detector Pump Model 400
2. Kitagawa Carbon Dioxide Detector tubes, No. 126a
3. Kitagawa Carbon Dioxide Concentration charts

Procedural considerations

This procedure will measure 0.1% to 2.6% quantities of CO_2 in the atmosphere. The *purple reagent* in the tube will change to pale pink, depending on the concentration of CO_2. Carbon dioxide neutralizes caustic soda to discolor phenolphthalein. Normal sampling time is 5 minutes with one pump stroke. The relative error will range from $\pm$ 10%. If a greater quantity of air is to be sampled, repeat the sampling procedure at 5-minute intervals and calculate the quantity of air tested against the reading on the concentration chart.

The following compounds react with the reagents in the tube accordingly; therefore, this exercise is also significant for the identification of the presence of these compounds in the atmosphere: Hydrogen chloride produces a similar pale pink stain. Hydrogen sulfide and/or sulfur dioxide produce a pale purple stain. Hydrogen cyanide, nitrogen dioxide, and chlorine produce a dark purple, purple, and white stain, respectively.

Procedure

1. Select four sampling areas, such as a factory area, power plant, shopping center, and campus.
2. Cut the tips off a detector tube by bending each tube end in the tube-tip breaker (Fig. 3, p. 12).
3. Insert the tube tip, marked with the red dot, securely into the pump inlet (Fig. 3).
4. Make certain that the pump handle is all the way in. Align the guide marks on the shaft and the back plate of the pump.
5. Pull the handle all the way out. Lock it with a 90-degree turn. Wait exactly 5 minutes.
6. Unlock the pump handle by making a

90-degree turn. If the pump handle retracts by more than 5 ml (¼ inch), a particle is probably lodged in the constant flow orifice. Repeat with a new tube.

7. After the 5-minute waiting period, remove the detector tube from the pump.

8. Arrange the tube vertically on the concentration chart with the *stained* (changed) end down.

9. Place the tube on the chart with the red dot at the top and the reagent between X and O.

10. Read the concentration of CO_2 (in %) indicated according to the length of the stain between X and O.

11. Post the %CO_2 ppm in the result table.

12. Collect the data from the other sampled areas and post in the result table.

13. Based on 0.5% to be equal to 5000 ppm, convert your reading to ppm and post in the result table.

14. Check the Introduction, p. 14, for data on carbon dioxide.

Result table

Sample	% CO_2 ppm	TLV	Difference	Remarks
1		0.5%		
2		0.5%		
3		0.5%		
4		0.5%		

NOTE: TLV = threshold limit value of 0.5% for 7 to 8 hours daily exposure at 25° C and 760 mm Hg for 1969.

Questions

Was there evidence of other gases present? See Procedural considerations.

What are some physiological effects on the respiratory system or skin of high concentrations of carbon dioxide?

Account for the cause of the differences in the carbon dioxide concentrations.

What are some major sources of carbon dioxide? How can they be controlled?

Is it safe to work or live in the areas in which the samples were taken?

Do high concentrations have an effect on the atmosphere? Explain.

Does carbon dioxide react with other compounds in the atmosphere and soil? Give examples.

Why should carbon dioxide be considered a pollution parameter?

5
Sulfur dioxide determination

Sulfur dioxide is emitted into our atmosphere by the burning of coal, oil, and gas. It is a colorless gas that reacts chemically to form many toxic compounds. It will react with the oxygen and hydrogen of the air to form sulfuric acid (H_2SO_4), which damages, among other things, our lungs. In concentrations of 0.01 to 0.02 ppm for long periods, it causes metal to corrode and interferes with pulmonary functions. At concentrations of 0.20 to 0.30 ppm for 2 to 4 days, it causes sore throat and eye irritation. At exposure for 3 days at 0.20 to 0.86 ppm, cardiorespiratory mortality increases. See Appendix, pp. 128 and 131.

Materials

The equipment and materials used for this test can be supplied by National Environmental Instruments, Inc., Fall River, Mass.

1. Unico Precision Gas Detector Pump Model 400 (one pump may be used by several students)
2. Kitagawa Sulfur Dioxide Detector tubes, No. 103d
3. Kitagawa Sulfur Dioxide Concentration charts

Procedural considerations

(Fig. 1.) This procedure will measure 1 to 80 ppm of sulfur dioxide in the atmosphere. The *purple* reagent in the tube will change to white depending on the concentration. The tetramethyldiaminodiphenylmethane is reduced and bleached by sulfur dioxide. Normal sampling time is 5 minutes. It is possible to detect as little as 0.5 ppm sulfur dioxide. This lower limit is not accurate but merely indicates a qualitative measurement. Refer to the temperature correction table if the temperature of the surroundings exceed or are lower than 20° C (68° F). One pump stroke is equivalent to 100 ml of air. The flow rate

is controlled by a restricting orifice that meters the air into the pump at a rate of 33 cc/minute.

Procedure

1. Select four sampling areas, such as a factory area, power plant, shopping center, and campus.
2. To sample, break the tips off a detector tube by inserting into the tube-tip breaker on the pump and twisting or bending the tube (Fig. 3, p. 12).
3. Insert the tube tip marked with the red dot securely into the pump inlet.
4. Make certain that the pump handle is all the way in. Align the guide marks on the shaft and the back plate of the pump.
5. Pull the handle all the way out. Lock it with a 90-degree turn. Wait exactly 5 minutes.
6. Unlock the pump handle by making a 90-degree turn. If the pump handle retracts by more than 5 ml (¼ inch), a particle is likely to be lodged in the constant-flow orifice. Repeat with a new tube if this happens.
7. After the 5-minute waiting period, remove the detector tube from the pump.
8. Arrange the tube vertically on the concentration chart with the *stained* (changed) end down.
9. Place the tube, on the chart, with the red dot at the top, and the reagent between X and O.
10. Read the concentration in ppm according to the length of the stain between X and O.
11. Post the ppm sulfur dioxide in the result table.
12. Collect the data for the other sample areas and post in the result table.
13. Check the Introduction and Appendix, p. 128, for data on sulfur dioxide.

Result table

Sample	ppm SO$_2$	TLV ppm	Difference	Remarks
1		5		
2		5		
3		5		
4		5		

NOTE: TLV = threshold limit value of 5 ppm for 8 hours daily exposure at 25° C and 760 mm Hg for 1968.

Questions

What are some physiological effects on the respiratory system, or skin, of high concentrations of sulfur dioxide?

Account for the cause of differences in sulfur dioxide concentration.

What are some major sources of sulfur dioxide? How can these be controlled?

Is it safe to work or live in the areas in which the samples were taken?

How do high concentrations of sulfur dioxide in the atmosphere effect the flora and fauna of the area?

Do these concentrations affect the soil? Give an example.

Why should sulfur dioxide be considered a pollution parameter?

6
Nitrogen dioxide determination

Materials

The equipment and materials used for this test can be supplied by National Environmental Instruments, Inc., Fall River, Mass.

1. Unico Precision Gas Detector Pump Model 400 (one pump may be used by several students)
2. Kitagawa Nitrogen Detector tubes, No. 117
3. Kitagawa Nitrogen Detector charts and temperature conversion table

Procedural considerations

(Fig. 3) This procedure will measure 1 to 1000 ppm nitrogen dioxide in the air. The white reagent will turn to a pale greenish orange depending on the concentration of NO_2. Nitrogen dioxide oxidizes o-tolidine to form nitroso-o,o-tolidine. The normal sampling time is 3 minutes. It is possible to detect levels as low as 0.1 ppm. This low concentration is not accurate but merely indicates a qualitative measurement. The concentration chart calibrations are based upon a tube temperature of 20° C (68° F), and normally correction is not necessary between 0° C (32° F) and 40° C (104° F). There are several elements and compounds that may interfere or react with the reagent in the tube. They are chlorine, bromine, and iodine, and chlorine oxide, which at concentrations of 1 ppm will produce a similar color. More than 5 ppm of ozone will also produce a similar color reaction. Nitrous oxide will discolor the whole reagent to yellow or pale green. Therefore, it is possible not only to determine the quantity of NO_2 in ambient air but also, under certain conditions, to identify the presence of the interfering compounds that are also contributors to air pollution.

Procedure

1. Select four sampling areas, such as a factory area or shopping center.
2. Cut off the tips of a fresh detector tube by inserting and twisting the tube ends into the tip cutter (Fig. 5).
3. Insert the tube tip marked with the red dot into the pump inlet.
4. Make certain the pump handle is all the way in. Align the guide marks on the shaft and the back plate of the pump.
5. Pull the pump handle all the way out. Lock it with a turn. Wait 3 minutes.
6. After 3 minutes unlock the pump handle. If the handle retracts more than 5 mm (1/4 inch), a particle may be lodged in the orifice. Repeat the procedure with a new tube.
7. Remove the detector tube from the pump inlet after completion of the sampling time.
8. Place the tube vertically on the concentration chart with the stained end down.
9. Position the top X and the bottom O boundaries between the reagent and end plugs on lines X - X and O - D respectively.
10. Read the concentration according to the length of stain O - C.
11. Post the ppm of nitrogen dioxide in the result table.
12. Collect the data from the other sampled areas and post.
13. Refer to the Appendix, pp. 129 and 131, for information on the effects of nitrogen dioxide.

Result table

Sample	ppm NO$_2$	TLV ppm	Difference	Remarks
1		5		
2		5		
3		5		
4		5		

NOTE: TLV = threshold limit value of 5 ppm (9 mg/m³) for a daily exposure of 8 hours.

Questions

What are some of the effects of high concentrations of nitrogen dioxide on the respiratory system?

What are some of the effects influencing infection?

What effect does photochemical oxidation have on itrogenous compounds?

What effect do nitrates have when they enter our blood system?

Does NO$_2$ have an effect on plants? Give examples and relate them to environmental quality.

How can the quantity of NO$_2$ be controlled?

PART II

Chemical analysis of water

Sampling considerations and water testing

TAKING WATER SAMPLES

Water samples should be drawn as closely as possible to the source of supply. Only carefully cleaned beakers, flasks, or bottles should be used for collection of samples. When well-water supplies are sampled from a pump, make certain that the pump has been running without interruption for a least 10 minutes before taking a sample.

A recommended sampling procedure is to extend a clean rubber hose to the bottom of the sample bottle, then to let the water overflow until the volume in the bottle is replaced three to four times and then to withdraw the rubber hose slowly while the water is still running. If this is not possible, fill it slowly with a gentle stream of water, taking care to avoid turbulence or air bubbles.

SAMPLE DILUTION TECHNIQUES
(choosing the proper size of water sample)

In some tests it may be found that the color that developed in the sample is too intense to be measured, and, in some tests, colors other than those expected occur. In both cases, it is necessary to dilute the original sample.

For example, when performing some tests, the colorimeter may read less than 10% transmittance (T). (For most accurate work it is desirable to use the range of 30% to 70%). The test must be repeated, but with the 25 ml graduated cylinder filled to the 12.5 ml mark with the water and then filled to the 25 ml mark with demineralized water. If 1 ppm is determined to be present and the sample was diluted to twice its original volume (12.5 to 25 ml), the 1 ppm reading should be multiplied by 2, which makes the ppm of the sample 2 ppm.

As an aid, the sample dilution table below shows the amount of water sample taken, the amount of demineralized water used to bring the volume up to 25 ml, and the multiplication factor.

A convenient way to carry out the sample dilution is to pipette the chosen sample portion into a clean graduated cylinder (or into a clean volumetric flask, for more accurate work) and then to fill the cylinder (or flask) to the desired volume with demineralized water.

Sample dilution table

Size of water sample taken	Demineralized water used to bring the volume to 25 ml	Multiplication factor
25 ml	0 ml	1
12.5 ml	12.5 ml	2
*10 ml	15 ml	2.5
* 5 ml	20 ml	5
* 2.5 ml	22.5 ml	10
* 1 ml	24 ml	25
* 0.25 ml	24.75 ml	100

*For sample sizes of 10 ml or less, a pipette should be used to measure the sample into the graduated cylinder or volumetric flask.

VOLUME MEASUREMENTS

When a graduated cylinder is filled to the mark with a water sample, the bottom of the meniscus should just touch the mark. All volume measurements are made by reading the bottom of the meniscus. Volumetric flasks, cylinders, and pipettes are filled until the bottom of the meniscus just touches the mark. This applies mainly to glassware. Plastic ware will cause little or no meniscus to be formed.

ACCURACY OF VOLUME MEASUREMENTS

In most of the procedures in this manual, graduated cylinders are used to measure the sample volumes, With few exceptions, this technique is accurate enough for the types of tests to be performed.

When the procedure calls for the addition of 1.00 ml of a reagent, it is recommended that a 1 ml transfer pipette be used.

When a sample must be diluted, it is strongly recommended that pipettes be used for the volume measurements. Pipettes are especially required for measuring small samples that are to be diluted, in order to minimize the percentage error.

TEMPERATURE

All of the tests outlined in this manual should be performed at a sample temperature between 20° C (68° F) and 27° C (80° F) to ensure accuracy. If certain tests require closer temperature control, it will be indicated in the procedural information.

POWDER PILLOW REAGENTS

To minimize problems of leakage and of deterioration, dry powdered reagents are used wherever possible. These reagents are conveniently packaged in individual, premeasured, polyethylene powder pillows, by Hach Chemical Co., Ames, Iowa. (Somewhat comparable materials are now marketed by LaMotte Chemical Products Co., Chestertown, Md.) Each pillow contains sufficient reagent for one test. The pillows are opened with clippers, scissors, or a knife.

UNTREATED WATER SAMPLE

In some test procedures, the term "untreated water" or "original water sample" appears as the standardization liquid. This means that the raw water is to be used without the addition of any chemicals.

REAGENT BLANK IN COLORIMETRIC TESTS

The term "reagent blank" refers to that effect wherein the reagent itself adds some color or turbidity to the sample being tested, thus giving erroneously high readings. In several of the tests, the blank is of such magnitude that it is compensated for, each time the test is performed, by standardizing the instrument on a blank sample such as in the case of ammonium nitrogen.

In most cases, however, the reagent blank is so very small that the instrument is standardized on an untreated portion of the original water sample, or on demineralized water. This is done routinely without any significant loss of accuracy, except when extremely small amounts of a constituent are sought. In this case, it is best for the analyst to determine the reagent blank. This is generally done by performing the test on a sample of high-quality demineralized water, which is also free of turbidity. The result is expressed as the ppm reagent blank and is substracted from the results of subsequent tests using that particular bottle of reagent. It is only necessary to determine the reagent blank when the material is first used, and then at intervals of several months, unless subsequent contamination is suspected.

Every effort should be made to purchase reagents with the lowest possible blank. They should be less than 0.2% of the full scale value (or about 2% on the percent transmittance meter scale). In some instances, it is either impossible or not practical to purchase reagents with this low a blank. In these cases, the only practical way is to determine the reagent blank as explained above and to subtract this from each determination.

INTERFERENCES

Many analytical procedures are subject to interference from substances that may be present in the sample. Most of the common interferences are mentioned in the procedures that follow, or in the notes accompanying the procedures. The elimination of many interferences are provided for in the reagent formulations themselves. Others may be eliminated by special sample pretreatments, which are described in the procedural information.

Even the presence of too much of the constitutent being tested for may constitute an interference. For example, the presence of a large excess of the parameter (say, 10 ppm) will cause the test to read less than full scale. As the sample is diluted down to 5 ppm, the test will then read higher than full scale, which indicates the need for further dilution, until the meter indication is "on scale."

One good general rule to observe is that when a result is suspected, (either by the fact that an unusual odor or turbidity is noticed), the test should be repeated on a sample diluted with demineralized water, (see Sample Dilution Techniques p. 21) and the result (corrected for the dilution) compared with the result of the original test. If these two results are not identical, the original result is probably in error and a further dilution should be made to check the second test (first dilution). This process is repeated until the same corrected result is obtained on two successive dilutions.

A more complete discussion of interferences and how to deal with them is found in the general introduction to *Standard Methods for the Examination of Water and Wastewater* (ed. 12, 1965, or 13, New York, 1971, American Public Health Association). The instructor is urged to obtain this book and to refer to it whenever problems are encountered.

STABILITY STANDARDS

Although the reagent formulations used in the tests described in this manual are, we believe, the most stable forms available on the market today, there is a certain finite useful lifetime for any chemical preparation. In view of this fact, it is well for the prudent analyst to occasionally test himself and his reagents by performing a

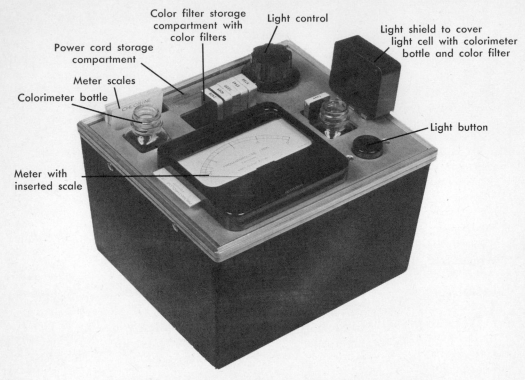

Color filter storage compartment with color filters

Light control

Light shield to cover light cell with colorimeter bottle and color filter

Power cord storage compartment

Meter scales

Colorimeter bottle

Light button

Meter with inserted scale

Fig. 4. Hach alternate current–direct reading (AC-DR) colorimeter.

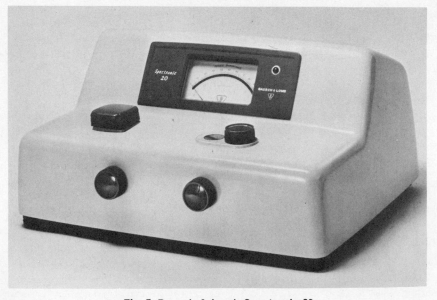

Fig. 5. Bausch & Lomb Spectronic 20.

test on a standard solution (solution of exact, known concentration of the substance being tested for). For example, to test the procedure and reagents in a phosphate test, a solution containing a known amount of phosphate is tested and the test result is compared with the known concentration. If the result compares favorably, one can continue to use the test with confidence. If the test result does not agree favorably with the known answer, then the reason should be determined. It may be caused by an incorrectly performed procedure, a deteriorated reagent, or even a defective standard solution, since these solutions too are subject to deterioration and should be prepared or otherwise obtained fresh whenever there is any doubt about their usefulness. Whenever doubt exists concerning the interpretation of a particular test of this type, check *Standard Methods for the Examination of Water and Wastewater*.

Some standards, such as chlorine, nitrite, and cyanide, are not stable. They must be prepared and used within a very short period of time. Directions for the preparation of such standards are available on request from the supplier.

Colorimeter or spectrophotometer calibration

Many types of spectrophotometers or colorimeters may be utilized in any of the following experiments. In reading the scales of the various instruments, somewhat different procedures must be followed in converting the scale units of the instrument into concentration units (ppm).

For the Hach Direct Reading Colorimeter (Fig. 4), with the appropriate filter in place, the scale directly indicates the concentration of the substance in question, with no conversions necessary.

For use with the Bausch & Lomb Spectronic 20 (Fig. 5), calibration charts have been prepared for each compound and are provided at the end of appropriate exercises. The concentration is indicated at the intersection of the left hand column and the upper row, the sum of which equals the percent transmittance (scale reading). For example, using the calibration chart on p. 29 for free chlorine, a Bausch & Lomb Spectronic 20 reading of 28% transmittance would give a concentration of 0.955 ppm.

If other spectrophotometers and colorimeters are employed, they must be calibrated with standard solutions by the procedure outlined below. Each exercise lists the wavelength in nanometers (nm) (same as millimicrons—mμ) to be used. When filters are used, consult your instrument manual, the manufacturer, or the manufacturer of the filters for their spectral response.

Procedure

1. Perform the test, using various known concentrations.
2. Adjust the instrument to full scale, using either the raw water or a reagent blank as specified by the instructions.
3. Measure the color of the various standards in percent transmittance.
4. Prepare a calibration curve by plotting the data on semilogarithmic graph paper. (Plot percent transmittance [% T] values on the logarithmic scale and concentration values on the linear scale.)
5. Connect the points with a smooth curve. See graphical representation of data, "General Introduction" in *Standard Methods for the Examination of Water and Wastewater* ed. 12 or 13 (see reference, p. 23).
6. It may be observed that additional points are required or questionable values exist. In this case repeat steps 1 to 3.
7. Values less than 10% T should not be included. Also remember that the most reliable portion of the scale is generally between 30% and 70% T.

7
Chlorine analysis

Chlorine as a gas, in the form of calcium hypochlorite or sodium hypochloride, is used as a disinfectant in the treatment of sewage, water, and industrial wastes. It serves as a means of reducing the population of pathogenic organisms to the level required by state and federal regulations. Chlorine is unique in that it reduces the virulence and viability of most microorganisms and macroorganisms. Pathogenic bacteria and potentially pathogenic bacteria such as those of the coliform group show a low tolerance to chlorine and are readily killed.

Chlorine is an extremely active chemical. It reacts readily with a large number of reducing materials of both organic and inorganic composition. The quantity of chlorine that reacts with these compounds, which may be found in the treated water, is referred to as the chlorine demand. The quantity of available chlorine that does not react with other compounds in the treated water is referred to as the residual chlorine. If the demand is great enough to combine with all the chlorine applied, there will be no residual; thus, there will be no active chlorine available to disinfect the water.

Most state, federal, and industrial facilities that are required to treat their wastes or water are required to show a chlorine residual of 0.5 ppm, 15 minutes after application. This residual is an indication of the disinfecting effect of the chlorine. Disinfection by chlorination must be a continuing process. Therefore, to provide the maximum sanitary and economic effect, it is necessary to continuously monitor the residual chlorine. This is usually accomplished through a visual colorimeter or by the use of a direct reading colorimeter. The so-called Black-Whittle method is used for both the total and free chlorine tests.

INSTRUCTOR: The reagents employed are those packaged by Hach Chemical Co., Ames, Iowa. Or refer to *Standard Methods for the Examination of Water and Wastewater,* ed. 12, 1965, p. 93, and ed. 13, New York, 1971, APHA.

Materials

1. Colorimeter or spectrophotometer
2. Free chlorine buffer, total chlorine buffer, chlorine indicator (*o*-tolidine)
3. Sample-collecting bottles, specimen droppers
4. Samples—four samples, three from different sources, including a sanitation treatment plant and one prepared by the instructor
5. 25 and 50 ml Erlenmeyer flasks

TEST 1—FREE CHLORINE TEST
Procedure

1. Fill in each identification space for the samples and number accordingly:

   1 _____.

   2 _____.

   3 _____.

   4 _____.

2. Add 25 ml of the test sample to clean colorimeter bottle or flask.
3. To the sample in the colorimeter bottle or flask, add 10 drops of free chlorine buffer; swirl to mix.
4. To the above add 10 drops of chlorine indicator solution and mix again.
5. *DR colorimeter*—Add 25 ml of the same test sample to another colorimeter bottle and place it in the light cell. Insert the appropriate free chlorine meter scale and

use the 4015 color filter. Adjust the light control to a meter reading of zero ppm. Then take this sample out of the light cell.

6. *DR colorimeter*—Put the prepared sample in the light cell. The ppm of free chlorine must be read within 5 minutes after the sample is prepared.

7. *B & L Spectronic 20*—Standardize with an untreated water sample. Wavelength—592 nm.

8. *B & L Spectronic 20*—Put the prepared sample in a ½-inch test tube. Insert it into the instrument and read the percent transmittance. Cross-refer T with the table on p. 29 to determine the ppm of free chlorine.

9. Repeat and record each of the test samples.

Result table

Samples	DR colorimeter ppm chlorine	Spectronic 20 % T	Spectronic 20 ppm chlorine
1			
2			
3			
4			

Question

What is the reason or reasons for the differences in ppm of free chlorine in each sample?

TEST 2—TOTAL CHLORINE TEST
Procedure

1. Using the test samples for test 1, add 25 ml to a colorimeter bottle.

2. To this, add 5 drops of the total chlorine buffer. Mix by swirling.

3. Add 5 drops of the chlorine indicator solution and mix. A violet color will appear if free or combined chlorine is present.

4. *DR colorimeter*—To another colorimeter bottle, add 25 ml of the sample and place it in the light cell. Insert the appropriate total chlorine scale and use the 4015 color filter. Adjust the light control for a meter reading of zero ppm. Then take the bottle out of the light cell.

5. *DR colorimeter*—Put the prepared sample in the light cell. Read the ppm of total chlorine immediately from the scale.

6. *B & L Spectronic 20*—Standardize with an untreated sample. Wavelength—592 nm.

7. *B & L Spectronic 20*—Put the treated sample into a ½-inch test tube and insert into the instrument. Read percent transmittance. Record and cross-refer the % T with the B & L Spectronic calibration for ppm of total chlorine (p. 30).

8. The ppm of combined chlorine is the difference between the ppm of free chlorine in test 1 and the ppm of total chlorine found in this test.

9. Repeat the procedure with each of the samples and record the data.

Result table

DR colorimeter

Samples	Test 1 Free chlorine	Test 2 Total chlorine	Combined
1			_____ ppm
2			_____ ppm
3			_____ ppm
4			_____ ppm

Result table

Spectronic 20

Samples	Test 1 Free chlorine		Test 2 Total chlorine		Combined
	% T	ppm	% T	ppm	ppm
1					
2					
3					
4					

B & L SPECTRONIC 20 CALIBRATIONS
Test 1—Free chlorine (ppm)
592 nm, ½-inch test tube

% T	0	1	2	3	4	5	6	7	8	9
20							1.01	0.985	0.955	0.930
30	0.900	0.880	0.850	0.830	0.810	0.785	0.770	0.750	0.730	0.710
40	0.685	0.670	0.650	0.635	0.620	0.600	0.590	0.570	0.555	0.535
50	0.520	0.510	0.495	0.480	0.470	0.455	0.440	0.430	0.415	0.400
60	0.390	0.375	0.360	0.350	0.335	0.320	0.310	0.300	0.290	0.275
70	0.265	0.255	0.245	0.235	0.225	0.215	0.205	0.195	0.185	0.175
80	0.165	0.160	0.150	0.140	0.130	0.120	0.115	0.105	0.095	0.090
90	0.080	0.070	0.060	0.050	0.045	0.035	0.030	0.025	0.015	0.010

B & L SPECTRONIC 20 CALIBRATIONS
Test 2—Total chlorine
592 nm, ½-inch test tube

% T	0	1	2	3	4	5	6	7	8	9
0								1.00	0.945	0.900
10	0.860	0.825	0.790	0.760	0.730	0.710	0.685	0.665	0.645	0.625
20	0.600	0.580	0.565	0.545	0.530	0.515	0.500	0.485	0.470	0.460
30	0.450	0.440	0.430	0.420	0.410	0.395	0.385	0.375	0.365	0.355
40	0.340	0.330	0.320	0.310	0.305	0.295	0.290	0.280	0.275	0.270
50	0.260	0.255	0.250	0.240	0.230	0.225	0.217	0.210	0.205	0.200
60	0.190	0.185	0.180	0.175	0.170	0.165	0.155	0.150	0.145	0.140
70	0.130	0.125	0.120	0.115	0.110	0.105	0.100	0.097	0.093	0.090
80	0.085	0.077	0.075	0.070	0.065	0.060	0.055	0.052	0.048	0.045
90	0.040	0.035	0.030	0.025	0.020	0.015	0.012	0.010	0.007	0.005

Questions

What is the reason or reasons for the differences found between samples? Comment on each sample.

What is the reason for the difference between the ppm of free chlorine and combined chlorine?

Define free chlorine and combined chlorine.

What is the significance of these two experiments? Of what value are they in the study of environmental science?

What significant conclusions can be drawn from this test?

8
Ammonium nitrogen

NESSLER'S METHOD

Ammonia is a product of the nitrogen cycle. The nitrogen cycle is a natural phenomenon of the ecology of the environment. The degradation of organic material and the assimilation of the products provide a source of nutrients in the food chain of plants and animals. Ammonia, therefore, is a basic product of the microbiological decay of dead plant and animal tissue and their waste products. Today, many forms of agriculture fertilizer contain free ammonia or in combination with other compounds.

Excess agricultural ammonia that may accumulate in raw water is largely responsible for the pollution of that water. This nutrient promotes the growth and propagation of unwanted bacteria and algae that eventually die and decompose in the water, forming organic wastes and putrefaction. The presence of ammonia nitrogen in raw water also serves as an indicator for domestic pollution. The presence of ammonia in drinking water reduces the effectiveness of chlorine treatment. Larger amounts of chlorine need to be applied to provide the necessary chlorine residual needed to disinfect the water.

INSTRUCTOR: The reagents employed are those packaged by Hach Chemical Co., Ames, Iowa. Or refer to *Standard Methods for the Examination of Water and Wastewater,* ed. 12, p. 193, and ed. 13, p. 226 (see reference, p. 23).

Materials

1. Colorimeter or spectrophotometer
2. Nessler's reagent, Rochelle salt solution
3. 4 Test-sample containers, 25 and 100 ml graduate cylinders, 50 ml flask
4. Samples from a stream, a pond, and tap water and one prepared by the instructor; label each appropriately

Procedural information

The following method is taken from *Standard Methods for the Examination of Water and Wastewater.* There are two methods given by the American Public Health Association for this test: one is a distillation method for the separation of ammonia from water; the other is the direct nesslerization method. The distillation method allows the concentration of small amounts of ammonia, in addition to the elimination of other compounds. The direct nesslerization method determines ammonium nitrogen as low as 0.02 ppm when using a photoelectric colorimeter. The direct nesslerization procedure is given here.

In addition to calcium and magnesium, iron and sulfide may interfere by causing the formation of turbidity with Nessler's reagent. Pretreatment with zinc sulfate and alkali may be used to eliminate interference from these sources. See *Standard Methods* for details. There are a number of rarely encountered compounds (mostly organics) that may interfere. Some of them are hydrazine, glycine, various aliphatic and aromatic amines, organic chloramines, acetone, aldehydes, and alcohols. Some symptoms may be a yellowish or greenish off-color or a turbidity. If these compounds are present, it may be necessary to distill the sample before the test is performed. See *Standard Methods* for details. For dilution of samples, see Sample Dilution Techniques, p. 21.

The results of this test are given in terms of ppm of ammonium nitrogen (N). To express as ppm of ammonia (NH_3), multiply the N value by 1.21. To express as ppm of ammonium (NH_4^+), multiply the N value by 1.29.

Procedure

1. Identify the samples and number accordingly.

   1 _____.

   2 _____.

   3 _____.

   4 _____.

2. Collect 25 ml of the test sample in a graduated cylinder. Pour the sample into a colorimeter bottle or flask. The temperature of the sample should be 20° C. Above 20° C the test results will be too high. Below this temperature the results will be too low. Do not use test samples with a water hardness of over 100 ppm. If it is necessary, add 1 drop of Rochelle salt solution, which will demineralize the water.

3. Collect 25 ml of *demineralized* water in a colorimeter bottle or flask.

4. To steps 2 and 3, add 1 ml of Nessler's reagent. Mix by swirling. If ammonium nitrogen is present, a yellow color will develop. Allow 10 minutes for color development.

5. *DR colorimeter*—Put the colorimeter bottle that contains the prepared sample of demineralized water in the light cell. Insert the appropriate ammonium nitrogen meter scale into the meter and use the 5543 color filter, 425 nm. Adjust the light control to give a meter reading of zero ppm.

6. *DR colorimeter*—Insert the colorimeter bottle that contains the prepared unknown sample into the light meter cell and read the ppm of ammonium nitrogen indicated.

7. *B & L Spectronic 20*—Standardize the instrument with the prepared sample of demineralized water.

8. *B & L Spectronic 20*—Put the prepared sample in the ½-inch test tube, insert it in the instrument, and record the reading. Determine the ppm of ammonium nitrogen (N) from the B & L Spectronic 20 calibration table at the end of the exercise.

9. Repeat the procedure with each of the test samples and record the data in the data result table.

Result table

DR colorimeter

Samples   ppm ammonium nitrogen

1 _____

2 _____

3 _____

4 _____

Result table

B & L Spectronic 20

Samples	% T	ppm N	Remarks
1			
2			
3			
4			

B & L SPECTRONIC 20 CALIBRATIONS
Nessler's method—ammonium nitrogen (ppm)
425 nm, ½-inch test tube

% T	0	1	2	3	4	5	6	7	8	9
30							2.52	2.46	2.43	2.35
40	2.29	2.23	2.18	2.13	2.07	2.02	1.97	1.93	1.89	1.84
50	1.80	1.75	1.70	1.66	1.62	1.58	1.54	1.50	1.46	1.43
60	1.39	1.35	1.31	1.28	1.25	1.21	1.17	1.15	1.11	1.07
70	1.04	1.01	0.98	0.95	0.92	0.89	0.86	0.83	0.80	0.77
80	0.75	0.72	0.69	0.66	0.64	0.61	0.58	0.56	0.53	0.50
90	0.48	0.45	0.43	0.40	0.37	0.35	0.30	0.25	0.21	0.15

Questions

Account for the presence of ammonium nitrogen and the variation between samples.

How does the presence of ammonium nitrogen affect the quality of our environment?

What recommendations could you make to reduce the amount of potential ammonia pollution?

9
Nitrate nitrogen

The presence of significant amounts of nitrates in water is evidence of a high rate of biological metabolic activity within the water system. Waste products that contain unusually large quantities of nitrates or nitrate compounds contribute extensively to the rapid growth of algae. This is followed by a decay of these organisms, which thus contributes to the unwanted organic pollution of our water supply. The polluted water may serve as a reservoir of nutrients for the propagation of potential pathogenic organisms or toxic compounds. It is not to be construed that nitrates are not unnecessary. They are an integral part of the nitrogen cycle necessary for the continuance of the natural ecological conditions of a habitat. *Nitrobacter* bacteria aid in the decomposition of organic matter and convert nitrite, under aerobic conditions, to nitrates. The nitrates are then eventually used in the synthesis of proteins by living organisms.

The leaching and running off of excess agricultural fertilizer also contributes to the nitrates found in raw water. If pregnant women drink water that contains excessive amounts of nitrates, the infants may be affected by methemoglobinemia, which is also known as a blue baby condition.

The following method gives both nitrate and nitrite nitrogen. To obtain nitrate only, one must also run the nitrite test. Subtract the nitrite nitrogen value from the quantity of nitrate shown in this test to give the nitrate present.

INSTRUCTOR: The reagents employed are those packaged by Hach Chemical Co. Or refer to *Standard Methods for the Examination of Water and Wastewater,* ed. 12, p. 395, and ed. 13, p. 458 (see reference, p. 23).

Materials

1. Colorimeter or spectrophotometer.
2. Sampling bottles, 25 and 50 ml Erlenmeyer flasks, 250 μl micropipette, colorimeter bottles, size-2 polyethylene stoppers, 25 and 50 ml. graduate cylinders, 1 ml. pipettes
3. Hach NitraVer IV powder pillows (cadmium reduction reagent), bromine water, phenol solution 3%. (Cadmium reaction method with diazotization, using 1-naphthylamine-sulfanilic acid)
4. Samples from a stream, a pond, tap water, or a river

Procedural information

This procedure is written for a sample dilution of 1:100, and the colorimeter scale is calculated for this dilution. If more sensitivity is wanted, make dilutions according to the Sample Dilution Techniques, p. 21, and apply the correction factor. Trace amounts can be determined without a dilution and the results divided by 100, giving a range of 0 to 1.5 ppm.

The results of this test are expressed in ppm of nitrate nitrogen (N). If ppm of nitrate (NO_3) is wanted, multiply the result by 4.4.

Strong oxidizing and reducing substances may interfere by causing *precipitation*. Cupric ions may cause low results. Do not collect the samples with sodium thiosulfate in the collection bottle. Clean the bottles thoroughly with a brush after each test.

The reagent blank, which is usually caused by some undissolved reagent, is primarily in the range of 2 to 4 ppm as nitrogen when 1:100 dilutions are used. The blank is only about 0.02 to 0.04 ppm as nitrogen when the trace procedure is followed. See Reagent

Blank in Colorimetric Tests, p. 23. Do not worry about black deposits that may settle in the bottle. If the nitrite level is high as compared to the nitrate, a pretreatment is necessary. This is done by adding bromine water, a drop at a time, to the sample until the yellow color of the bromine water persists. After 3 minutes, add one drop of 3% phenol solution and proceed with the test.

Procedure

1. Identify and number the samples accordingly

   1 _____.

   2 _____.

   3 _____.

   4 _____.

2. Thoroughly rinse two colorimeter bottles or Erlenmeyer flasks and their stoppers three or four times with demineralized water.

3. Measure two samples of demineralized water by filling a 25 ml graduated cylinder to the 24.5 level. Pour the demineralized water into each rinsed colorimeter bottle or Erlenmeyer flask.

4. Using a micropipette, measure 0.25 ml (250 μl or 250 λ [lambdas]) of water sample and add it to one bottle or flask.

5. Add the contents of one NitraVer IV powder pillow to one bottle or flask, stopper, and shake for 1 minute. If nitrate or nitrite is present, a red color will develop. Wait 3 more minutes.

6. *DR colorimeter*—Put the other prepared colorimeter bottle with the treated demineralized water in the light cell. Insert the nitrate nitrogen meter scale in the meter. Use the 4445 color filter (525 nm). Adjust the light control for a meter reading of zero ppm.

7. *DR colorimeter*—As soon as the color has developed in the treated sample, place it into the colorimeter light cell and read the ppm of nitrogen that is present as nitrate and/or nitrite. Record in the result table.

8. *B & L Spectronic 20*—After step 5 is completed for the water sample, fill a 1/2-inch test tube with demineralized water, place it into the light cell. Standardize the instrument.

9. *B & L Spectronic 20*—When the color has fully developed in the flask with the treated sample, pour the 1/2-inch test tube about two thirds full. Measure the % T and determine the ppm of nitrate and/or nitrite nitrogen by cross referencing the reading with the Spectronic 20 calibration table that follows the questions. Post your results in the result table.

10. Refer to the Introduction and also the Appendix, p. 133, for additional information on nitrate nitrogen.

Result table

DR colorimeter

Samples   ppm nitrate nitrogen

1 _____

2 _____

3 _____

4 _____

Result table

B & L Spectronic 20

Samples	% T	ppm NO$_3$	Remarks
1			
2			
3			
4			

B & L SPECTRONIC 20 CALIBRATIONS
Cadmium reduction method—nitrate nitrogen (ppm)
525 nm, ½-inch test tube

% T	0	1	2	3	4	5	6	7	8	9
0										150
10	142	134	128	122	116	111	107	103	99	95
20	92	89	87	84	82	79	77	75	73	71
30	68	67	65	63	62	60	58	57	55	54
40	52	51	49	48	47	45	44	43	42	40
50	39	38	37	36	35	34	33	32	30	29
60	28	27	26	25	25	24	23	23	22	22
70	21	21	20	20	19	18	18	17	17	16
80	16	15	15	14	13	12	12	11	11	10
90	10	9	8	7	6	5	4	3	2	0

Questions

List some causes for the possible differences found in the amount of nitrogen in the test samples?

Is it necessary that nitrate nitrogen be present in our environment? Explain.

What affect does high concentrations of nitrates in the water supply have on humans?

Does a high concentration of nitrates affect the population of aquatic organisms (plants, algae, animals)? Explain.

Why is a limited quantity of nitrate necessary in our environment?

10
Nitrite nitrogen

Nitrite nitrogen is an intermediate product of the nitrogen cycle. It may occur in water as the result of biological decomposition of the organic nitrogen of dead plant and animal tissue. It is the product of the action of *Nitrosomonas* (bacteria), which convert ammonia to nitrites under aerobic conditions. Some bacteria also reduce nitrates to nitrites under anaerobic conditions. Nitrites are not normally present in raw water in large quantities, because of their ability to react chemically with other compounds or elements. However, the presence of nitrites in large quantity serves as an indicator of bacterial and organic pollution.

INSTRUCTOR: The reagents employed are those packaged by Hach Chemical Co., Ames, Iowa. Or refer to *Standard Methods for the Examination of Water and Wastewater,* ed. 12, p. 400, and ed. 13, p. 240.

Materials

1. Colorimeter or spectrophotometer
2. Graduate cylinders, 25 and 100 ml
3. NitriVer powder (direct diazotization, using 1-naphthylamine-sulfanilic acid)
4. Four samples, including one prepared by the instructor; use streams, lakes, ponds, or sewage effluent for test samples

Procedural information

For dilution samples, see Sample Dilution Techniques, p. 21. The results of this test are expressed in ppm of nitrite nitrogen (N). If it is desired to express the results in ppm of nitrite (NO_2), multiply the result by 3.3. Strong oxidizing and reducing substances interfere. Ferric, mercurous, silver, bismuth, antimonious, lead, auric, chloroplatinate, and metavanadate ions interfere by causing precipitation. Cupric ions may cause low results.

Very high levels of nitrate appear to undergo a slight amount of reduction to nitrite, either spontaneously, or during the course of the test, even with the use of NitriVer. Thus, if very high levels (such as 100 ppm) of nitrate are present, a slight amount of nitrite can be expected to be found also.

Procedure

1. Identify the samples in the spaces below and number accordingly:

   1 _____.

   2 _____.

   3 _____.

   4 _____.

2. Collect 25 ml of the test water sample in a graduate cylinder. Pour the sample into a colorimeter bottle or a 25 ml Erlenmeyer flask.
3. Add the NitriVer powder pillow. Shake to mix. Allow 15 minutes for the color to develop.
4. *DR colorimeter*—Put 25 ml of the sample water into another colorimeter bottle and insert into the light cell. Insert the recommended nitrite nitrogen meter scale into the meter and use the appropriate color filter (colorimeter filter No. 4445). Adjust the meter reading to zero ppm.
5. *DR colorimeter*—Place the prepared test sample into the light cell and read the ppm of nitrite nitrogen. Post the reading in the result table.
6. *B & L Spectronic 20*—After step 3, add water from the original sample to the test tube. Standardize the instrument (525 nm).
7. *B & L Spectronic 20*—To a test tube add the treated sample. Insert into the instrument. Measure the % T (525 nm) and

post it in the result table. Using the B & L Spectronic calibration table found after the questions, determine the ppm of nitrite nitrogen (N) and post your results.

8. Collect the data on all samples and record in the result table.
9. Refer to the Introduction and Appendix, p. 133, for information on nitrites.

Result table

DR colorimeter

Samples   ppm N

1 _____

2 _____

3 _____

4 _____

Result table

B & L Spectronic 20

Samples	% T	ppm N	Remarks

1 _____

2 _____

3 _____

4 _____

B & L SPECTRONIC 20 CALIBRATION
Diazotization method—nitrite nitrogen (ppm)
525 nm, ½-inch test tube

% T	0	1	2	3	4	5	6	7	8	9
10						0.35	0.33	0.31	0.30	0.285
20	0.27	0.26	0.25	0.243	0.235	0.226	0.20	0.21	0.205	0.20
30	0.19	0.185	0.18	0.173	0.17	0.163	0.16	0.153	0.15	0.144
40	0.14	0.136	0.132	0.128	0.124	0.121	0.118	0.114	0.111	0.108
50	0.104	0.101	0.098	0.095	0.092	0.089	0.086	0.083	0.08	0.078
60	0.074	0.072	0.07	0.067	0.064	0.061	0.059	0.056	0.053	0.052
70	0.05	0.049	0.047	0.045	0.043	0.041	0.04	0.039	0.037	0.035
80	0.033	0.032	0.031	0.029	0.028	0.026	0.024	0.022	0.02	0.019
90	0.018	0.015	0.013	0.011	0.01	0.008	0.006	0.005	0.003	0.001

Questions

What are the major sources of nitrite nitrogen?

At what level of concentration of nitrite nitrogen in raw water is it considered a pollutant? Why?

How do you account for the difference in the quantity of nitrite nitrogen in the test samples?

Why is nitrite nitrogen considered a water pollutant indicator?

Why is nitrite nitrogen considered to be an air pollutant? (Refer to Appendix, p. 133.)

11
Hydrogen sulfide

METHYLENE BLUE METHOD

Sulfides occur in soluble and insoluble forms. This exercise is for the determination of the total sulfide (soluble plus insoluble).

INSTRUCTOR: The reagents employed are those packaged by Hach Chemical Co., Ames, Iowa. Or refer to *Standard Methods for the Examination of Water and Wastewater,* ed. 12, p. 429, or ed. 13.

Materials

1. Colorimeter or spectrophotometer
2. 25 and 50 ml graduate cylinders, colorimeter bottles, ½-inch test tubes, 1 ml disposable pipettes
3. Sulfide I solution (acid, para-amino-dimethylaniline solution), Sulfide II solution (potassium dichromate solution), bromide water, phenol solution
4. Samples collected from four water sources, such as a well, pond, river, or stream

Procedural information

When the total sulfide is determined in a turbid water sample, do not use demineralized water as outlined in procedure 2, 7, and 8. Use a treated sample of which the sulfide has been removed for these procedures. To remove the sulfide from the turbid water sample, add bromine water, a drop at a time, to the sample until a yellow color that is stable appears. Then add a drop of 3% phenol to remove the excess bromine. If the results of this test are unsatisfactory, refer to *Standard Methods for the Examination of Water and Wastewater,* ed. 12, p. 429, or ed. 13. It is *not necessary* to treat the sample when using a sample that is not highly turbid.

Procedure

1. Identify the samples according to number.

   1 _____.

   2 _____.

   3 _____.

   4 _____.

2. Using a 25 or 50 ml graduate cylinder, measure 25 ml of demineralized water and pour it into a colorimeter bottle or Erlenmeyer flask.
3. Add 25 ml of the sample water to another colorimeter bottle or Erlenmeyer flask.
4. Pipette 1 ml of Sulfide I solution into each of the colorimeter bottles or Erlenmeyer flasks.
5. Pipette 1 ml of Sulfide II solution into each of the colorimeter bottles or Erlenmeyer flasks. Swirl to mix.
6. The solution will develop a pink color that will turn blue if sulfide is present. Allow 5 minutes for the color to develop.
7. *DR colorimeter*—Put the colorimeter bottle containing the prepared demineralized water sample into the light cell of the DR colorimeter. Insert the hydrogen sulfide scale using the 2408 color filter. Adjust the light control for a meter reading of zero ppm.
8. *DR colorimeter*—Place the colorimeter bottle containing the prepared water sample into the light cell and read the total sulfide in ppm of hydrogen sulfide. Post in the result table.
9. *B & L Spectronic 20*—Wavelength, 665 nm. Using the prepared demineralized sample, standardize.

10. *B & L Spectronic 20*—Insert the prepared water sample and read the % T. Determine the ppm of hydrogen sulfide from the table following the result table.

If the % T is less than 17, dilute the sample with demineralized water and multiply the percentage by the dilution factor. Post in the result table.

Result table

Sample	DR colorimeter ppm H_2S	Spectronic 20 % T	Spectronic 20 ppm H_2S	Remarks
1				
2				
3				
4				

B & L SPECTRONIC 20 CALIBRATIONS
Hydrogen sulfide (ppm)
665 nm, ½-inch test tube

	0	1	2	3	4	5	6	7	8	9
10								1.53	1.48	1.43
20	1.39	1.36	1.32	1.27	1.24	1.20	1.17	1.13	1.10	1.07
30	1.05	1.02	0.99	0.97	0.94	0.92	0.89	0.87	0.85	0.82
40	0.80	0.78	0.76	0.74	0.72	0.70	0.68	0.67	0.65	0.63
50	0.61	0.59	0.57	0.56	0.55	0.53	0.52	0.50	0.48	0.47
60	0.45	0.44	0.43	0.42	0.40	0.39	0.38	0.37	0.35	0.34
70	0.33	0.32	0.30	0.29	0.28	0.27	0.26	0.25	0.23	0.22
80	0.21	0.20	0.19	0.18	0.17	0.16	0.15	0.14	0.13	0.12
90	0.11	0.10	0.09	0.08	0.07	0.06	0.05	0.03	0.02	0.01

Questions

What are some physiological effects of high concentrations of hydrogen sulfide?

Account for the different quantities of hydrogen sulfide in the samples tested.

41

What are the major sources of hydrogen sulfide? How can they be controlled?

What effect, if any, do the concentrations of hydrogen sulfide in the water sources have upon the plant and animal inhabitants?

Hypothesize what different chemical compounds may be formed from hydrogen sulfide in the presence of water or soil. Which of these may be toxic to man?

Why should hydrogen sulfide be considered a pollution parameter that affects the qaulity of our environment?

12
Total phosphate (ortho and meta)

The natural water of lakes, rivers, streams, and other raw water sources normally contain low levels of phosphate in the range of 0.01 to 0.5 mg/L (ppm). This phosphate is in a soluble form. Phosphorus is considered a pollution indicator for a number of reasons: excess amounts in natural water promote the growth of algae and microorganisms, thus producing a large amount of unwanted organic matter. The decaying organic matter contributes to the biological oxygen demand (BOD), thereby oxidizing (reducing the quantity of) the dissolved oxygen that is necessary for the survival of beneficial aquatic organisms. It exists in several chemical forms that are highly reactive and unstable. Therefore, they are hard to eliminate in water purification processes.

There are many major contributors of phosphorus to our water environment. If left unchecked, these sources could cause the complete degradation in the quality of our water supply. The following are but a few of the sources of phosphorus: Run-off from agriculture fertilizers, human, animal, and plant residues and wastes, detergents in household wastewater, detergent wastes from industrial factory and detergent manufactures, chemical processing plants, and improperly managed municipal sewage plants. It is not the objective of this test to determine the quality and quantity of all the forms of phosphorus. This test is designed to give an estimated amount of total phosphate found in the water samples. Significant differences in the samples from different sources should be noted. Refer to the Appendix, p. 133, for additional information.

INSTRUCTOR: The reagents employed are those packaged by Hach Chemical Co., Ames, Iowa. Or, to prepare the reagents, refer to *Standard Methods for the Examination of Water and Wastewater,* ed. 12, p. 231, or ed. 13, pp. 530 to 532.

Materials

1. Colorimeter or spectrophotometer
2. Graduated cylinders, 25 and 100 ml; 250 ml Erlenmeyer flask; hot plate
3. Ammonium molybdate solution, amino acid solution or amino acid powder pillows (amino-naphthol-sulfonic acid method)
4. Four test samples from a lake, a stream, a pond, and a sewage facility

Procedural information

If the original water sample is clear, it may be used as the blank sample for standardizing the instrument in step 8 or step 10, without the addition of the ammonium molybdate as directed in step 4. In all cases, but especially when the concentration of phosphate is greater than 15 ppm, EXTREME care must be taken to see that the proper temperature and color development time are maintained. For example, if the sample temperature is 28° C during the 10-minute color development period, a 25 ppm sample will read greater than 35 ppm. At 20° C, this 25 ppm solution would read about 20 ppm. This same variation is observed when the color development time is varied from the prescribed 10 minutes. (For the B & L Spectronic 20, the IP40 photocell and the red filter must be used in place of the 5581 photo-cell in this test.)

Sulfide interferes by forming a blue color directly with the molybdate reagent. Sulfide is removed by preliminary oxidation with permanganate or bromine water. Nitrite bleaches the blue color. Nitrite may be removed by preliminary treatment with sulfamic acid. When phosphate is determined in waters containing high salt levels, low results may be obtained. If salt water is encountered, dilutions should be made and tested until two successive dilutions yield essentially the same results. See Sample Dilution Techniques, p. 21. If a color other than blue is formed, a sample dilution is required. As the concentration of phosphate increases, the color changes from *blue* to *green,* then to *yellow,* and finally to *brown.* The *brown* color may indicate a concentration as high as 100,000 ppm PO_4. If dilution is necessary, see Sample Dilution Techniques. For wastewater, it is recommended that the sample not be filtered to remove turbidity. Instead, dilute the sample 9 : 1 to reduce turbidity to an insignificant level and to bring the phosphate content down to an easily analyzed level.

Procedure

1. Identify and number the samples.

    1 _____.

    2 _____.

    3 _____.

    4 _____.

2. Collect 25 ml of the test sample and pour it into a 250 ml flask.

3. Collect 25 ml of the test sample and pour it into a colorimeter bottle or 25 ml Erlenmeyer flask.

4. Put 1 ml of ammonium molybdate solution into the 250 ml flask and colorimeter bottle or flask of steps 2 and 3. Swirl to mix.

5. Gently boil the contents of the 250 ml flask for 10 minutes or in boiling water bath and then cool to 24° C. (This procedure should be done under an exhaust hood)

6. Transfer the boiled test water sample 5 to a graduated cylinder and bring the volume to 25 ml with distilled water. Then put into a colorimeter bottle or 25 ml Erlenmeyer flask.

7. To the colorimeter bottle or flask containing the boiled water sample, add 1 ml of amino acid solution or one amino acid powder pillow and mix. If phosphate is present, a blue color will develop to the maximum in 10 minutes.

8. *DR colorimeter*—Put the colorimeter bottle prepared in steps 3 and 4 (not boiled) into the light cell. Insert the high- or low-range phosphate meter scale (amino acid method) into the meter and use the recommended filters. Adjust the meter to a reading of zero ppm.

9. *DR colorimeter*—Place the colorimeter bottle with the prepared (boiled) sample into the light cell and read the ppm of total phosphate.

10. *B & L Spectronic 20*—Use the prepared sample of steps 3 and 4 to standardize the instrument.

11. *B & L Spectronic 20*—After step 7, pour the prepared sample into the ½-inch test tube. Using 710 nm, measure the color and determine the ppm (ortho- and meta-)phosphate from the B & L Spectronic calibration table.

12. *B & L Spectronic 20*—Post the ppm of phosphate in the result table. Collect the data for the other samples and also post in the table.

13. Collect the data for all samples and post the results in the result table.

14. Refer to the Introduction and Appendix, p. 133, for additional information on phosphate.

Result table

DR colorimeter

Samples	Total phosphate ppm	Recommended level (state or federal)	Difference
1			
2			
3			
4			

Result table

B & L Spectronic 20

Samples	% T	ppm phosphate	Recommended level (state or federal)	Difference
1				
2				
3				
4				

B & L SPECTRONIC 20 CALIBRATIONS
Amino acid method—total phosphate (ppm)
710 nm, ½-inch test tube

	0	1	2	3	4	5	6	7	8	9
20		30.0	27.1	25.0	23.3	22.9	20.5	19.5	18.5	17.6
30	16.7	15.8	15.0	14.3	13.5	12.9	12.4	11.7	11.2	10.5
40	10.3	10.0	9.5	9.2	8.8	8.5	8.2	8.0	7.7	7.5
50	7.2	7.0	6.8	6.6	6.4	6.2	6.0	5.8	5.6	5.4
60	5.2	5.1	4.9	4.7	4.6	4.4	4.2	4.1	3.9	3.7
70	3.6	3.5	3.3	3.1	3.0	2.8	2.7	2.6	2.4	2.3
80	2.2	2.0	1.9	1.8	1.6	1.5	1.4	1.3	1.2	1.0
90	0.9	0.9	0.8	0.7	0.6	0.5	0.4	0.3	0.2	0.1

Questions

How can an excess of phosphorus compounds in natural water affect the ecology of our environment?

Formulate a list of possible sources of the phosphate found in the samples.

Could the phosphate present in each of the sample sources influence the aquatic organisms that live in the area? Why or how?

13
Determination of mercury in water

Mercury is a metal that is used extensively in industry and laboratory research. Many compounds formulated from mercury have detrimental effects on our environmental quality. The compounds that are of direct concern are the pesticides, mercury cells for caustic chlorine production, dental preparations, fungicidal and bactericidal additives to paints, batteries, and catalytic compounds. Other dangerous compounds, the mercuric salts, are extensively produced; some examples are mercuric chloride ($HgCl_2$, corrosive sublimate), a violent poison; mercurous chloride ($HgCl$), sometimes used in medicinal preparations; and mercuric sulfide vermilion, used by the paint industry. Methyl mercury produced in the environment and by bacteria interferes with the nervous system of man and animals. Considerable methyl mercury is produced in the synthesis of acetaldehyde, a compound used extensively in laboratory research. Only small amounts of mercury are eliminated at a time by humans. Therefore, since it accumulates in the body, low levels of exposure or consumption may eventually cause violent death. Mercury, a volatile element, readily vaporizes in air and is a virulent poison in its elemental form. It is readily absorbed through the respiratory tract, through the broken skin, and into the gastrointestinal tract. The maximum allowable concentration of mercury vapor in air has been set at 0.1 mg/m.^{3} As temperature increases, the toxicity absorbancy of mercury also increases.

Many marine animals accumulate as much as 0.03 mg/L (0.03 ppm) of mercury in their tissue. It is normally present in ocean water at levels of 0.00003 to 0.00004 mg/L. In some areas of the world, such as Japan, people have been affected with severe neurological disorders as the result of eating ocean fish, crabs, and oysters. Deaths of many water fowl have been traced to methyl mercury compounds. Typical symptoms of lethal levels of mercury are cerebellar ataxia, constriction of vision fields, and dysarthria (imperfect articulation of speech). There are also direct effects on the physiological activities of the cerebellum and cerebral cortices. Concentrations as low as 0.1 mg/L have been found to inhibit the photosynthesis processes of aquatic plants. The half-life of mercury in humans is about 2 years.

MERCURY IN WATER (DITHIZON METHOD)

INSTRUCTOR: The materials and procedures used in this exercise are those recommended by Hach Chemical Co., Ames, Iowa.

Materials

1. Colorimeter
2. Reagent dispenser, extractor colorimeter cell, interference color filter for 610 nm
3. Chloroform ACS, MercuVer I powder pillows (alkaline permanganate and chlorine—strong oxidizing agent), MercuVer II powder pillows (dithizon and carrier), MercuVer III powder pillows (hydroxylammonium sulfate—strong reducing agent), standard mercuric chloride 12.5 ppm Hg^{++}, sulfuric acid 14.5 N
4. Gelman filter, 47 mm, Type E, No. 61631 or regular filter paper, Millipore filter holder
5. 100 and 500 ml graduate cylinders, hot plate

Procedural information

The dithizon method for the determination of the presence of mercury in water has been adapted for use with the DR colorimeter. As little as 0.003 ppm of mercury can be determined by this method. The procedure is based upon a titration using a solid granular dithizon reagent (MercuVer II). The Hach DR colorimeter is used to indicate the equivalence points. When incremental additions of dithizon are added to the water sample, the colorimeter does not register a reading until a surplus of dithizon is reached.

Four samples should be tested. Students can work in groups of four. Each student should do one test and collect the data of the other students. Then post the results on the graph and table for comparison.

TEST 1—WATER WITHOUT ORGANIC MATTER
Procedure

1. Identify the water samples:

   1 _____.   3 _____.

   2 _____.   4 _____.

2. Put 250 ml of the water sample into the extraction colorimeter cell.
3. Add 22.5 ml of 14.5 N sulfuric acid to that of step 2.
4. Add 25 ml of chloroform. Put on the stopper and shake for 5 seconds. Release

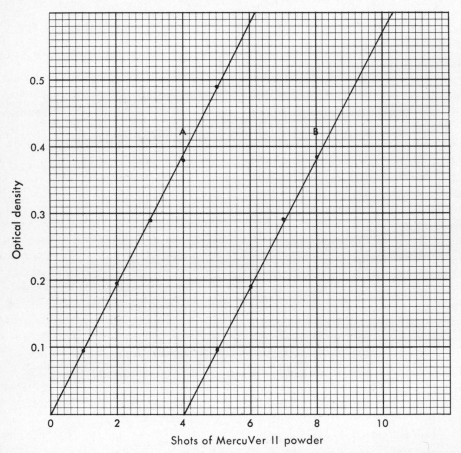

Graphic presentation. Curve **A** was obtained when analyzing mercury-free water. The curve intersects zero. Curve **B** was obtained when analyzing a standard sample containing 0.04 ppm mercury. After the addition of four shots of dithizon reagent, the optical density remained zero. When five or more shots were added, the result was an increase in optical density because of the presence of excess green dithizon in the sample. Projecting the curve plotted through the series of points beyond four, produced a straight line that intersected the axis at 4. Thus, four shots of dithizon were equivalent to 0.04 ppm mercury, or one shot was equivalent to 0.01 ppm mercury.

the air pressure by loosening the stopper. (CAUTION: Do not inhale the fumes or spill the sample; perform this step under an exhaust hood.)

5. Insert the 610 nm color filter into the slot in the colorimeter. Place the extractor into the colorimeter cell holder. Adjust for an optical density reading of zero. (Do not locate the DR colorimeter directly under overhead light. Place in a subdued light area.)

6. With the dispenser, add one shot of MercuVer II powder, cap the extraction cell, and shake for 5 seconds. Put the cell back into the colorimeter and wait for the chloroform to settle.

7. Read and record the optical density on the graph.

8. Repeat steps 6 and 7 at least four times. Record the optical density each time. On the graph paper draw a straight line that intersects with each optical density plotted. (Refer to the example graphic presentation.)

Determination of the results

1. Plot the results on the graph paper and extend the best straight line to the axis. (See graphic presentation.) Read the equivalent number of shots of MercuVer II from the scale and multiply by the standardization factor for the particular lot of MercuVer II used. See 2 below.

2. Calculations to determine ppm of mercury present in the sample:
   a. Determine the number of shots of MercuVer II at the equivalence point. Call this number "E."
   b. Determine the increase in optical density *per* shot of MercuVer II in the region *beyond* the equivalence point. It should be within the range of 0.05 to 0.1. (If it is less, replace the MercuVer II.) Call this number "D."

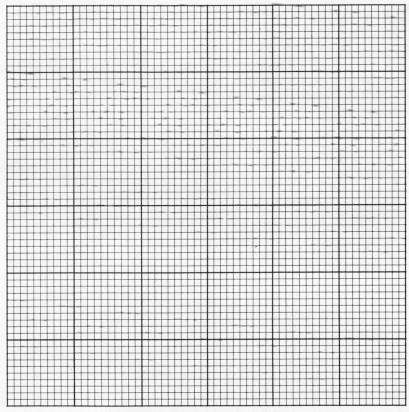

Graphic presentation to be filled in

c. Determine the ppm of mercury present in the sample with the formula:
$$ppm = E \times D \times 0.13.$$
3. Determine the ppm of mercury in your sample. Post in the result table. Collect the results from the other three samples and post in your result table.
4. For information on the recommended concentrations or limits, refer to the Introduction to this exercise.

Result table

Samples	ppm mercury	Dangerous level (yes or no)
1_____		
2_____		
3_____		
4_____		

Questions

What is a possible source of the mercury found in each sample?

Would the levels of mercury be considered lethal to plants, animals, and humans?

How could the presence of mercury be controlled in our environment?

If a sample is a source of drinking water, is it toxic or will it become lethal to humans? Support your answer.

TEST 2—MERCURY TEST IN THE PRESENCE OF ORGANIC MATTER
Procedural information

This procedure has been developed to oxidize the organic matter with alkaline permanganate and chlorine. It is effective for testing raw domestic sewage or a bottom sample that contains organic matter.

Procedure

1. Identify the four samples collected.

1 _____.   3 _____.

2 _____.   4 _____.

2. Measure a 250 ml sample into a 500 ml flask.

3. Add the contents of one MercuVer I powder pillow.
4. Heat the sample to a boil until most of the organic matter is destroyed.
5. Remove the sample from the hot plate and, after boiling has stopped, add 25 ml of 14.5 N sulfuric acid. (Do not inhale the fumes.) Return the sample to the hot plate and boil gently for 2 minutes. (Complete under an exhaust hood.)
6. Remove the sample from the hot plate and allow boiling to stop. Add the contents of one pillow of MercuVer III.
7. Return the sample to the hot plate and boil gently for approximately 1 minute. Cool to room temperature (a water bath may be used).
8. Filter the sample through fluted filter paper or the Gelman fiber glass filter pad using a Millipore filter holder and funnel.
9. Proceed with steps 4 through 8 of the procedure on p. 48.
10. Apply the same methods for determining the ppm of mercury that are outlined under the subject of determination of results on p. 49.

Result table

Samples	ppm mercury	Dangerous level (yes or no)
1		
2		
3		
4		

Questions

What material in each sample do you feel contained the most mercury?

Name several possible sources of the mercury.

Support your answers on the dangerous levels.

How can the quantity of mercury be controlled in our environment?

What are the EPA (Environmental Protection Agency) recommended limits for mercury in food and water?

14
pH determination

The determination of the pH of a solution of substance provides a means for the measurement of the acidity or alkalinity of that material. It is based upon the negative logarithm of the hydrogen ion present in the solution. It is used extensively in research and as a water quality indicator in water analysis. The pH units are measured in increments from 0 to 14. Readings from 0 to 7 indicate a decreasing acid condition. A pH of 7 represents a neutral condition. Readings beyond pH 7 through 14 indicate an increasing alkalinity condition. The pH of natural water normally ranges from 5 to 8. However, most potable drinking water ranges between 6.5 to 7.5.

Research data has shown that the acidity and alkalinity of the water and soil in some areas of the United States is changing. This has been brought about by the emission of acid-forming and highly alkaline waste compounds into the air, rivers, and lakes by processing and chemical plants. The salting of highways and city streets during the winter months adds thousands of tons of salts to relatively small areas. These salts dissolve in the run-off water and eventually reach our water supply in high concentrations. Subsequently, the high concentration of salts can have a direct effect on the alkalinity of the raw water. It can be said, that the population of aquatic flora and fauna and the water quality of our environment are both directly and indirectly influenced by the pH of the water environment.

TEST 1—COLORIMETER METHOD
Materials

1. Colorimeter or spectrophotometer
2. Graduate cylinders and flasks, 25 and 100 ml

3. Appropriate standard indicator, wide range–indicator pH solutions (4.0 to 10.0)
4. Four samples from a lake, stream, pond, and tap water and prepared by the instructor

Procedural information

If chlorine is present, remove by using standard dechlorination procedure in *Standard Methods for the Examination of Water and Wastewater* for chlorine removal procedure.

Procedure

1. Identify and number the samples.

   1 _____.

   2 _____.

   3 _____.

   4 _____.

2. Measure 25 ml of the test sample into a graduate cylinder.
3. Add 1 ml of the appropriate standard indicator. Stopper and shake. Transfer the contents to a colorimeter bottle.
4. Fill a colorimeter bottle with the untreated test sample and insert into the light cell. Insert the meter scale, No. 2857, and the recommended color filter, 4084. Standardize by adjusting the light control so that the meter reads at the far right end of the arc. For a spectrophotometer, use 525 nm.
5. Insert the prepared sample into the light cell and read the pH.
6. Repeat the procedure, if the reading falls outside either end of the scale by using a

different indicator. However, the wide range should serve most purposes.

7. Repeat the procedure with each test sample and post results in the data record.

8. Check your results with recommended concentrations or tolerances found in the Introduction to this exercise.

Result table

Sample	pH	Acid or alkaline
1		
2		
3		
4		

Questions

Which of the samples indicate that the source is acid, alkaline?

Formulate possible causes of the acid or alkaline condition of the sources.

Can you expect to find a difference caused by the pH in the flora and fauna present in the sources? Explain.

What is the effect of pH on our total environmental quality?

TEST 2—HYDRION PAPER PROCEDURE

INSTRUCTOR: Use the Hydrion paper procedure for a quick semiqualitative method to determine the acidity or alkalinity of the water sample. Demonstrate the procedure to the students. Instruct them to post the determinations on the result table of Test 1.

15
Anionic surfactants (detergents)

The dictionary defines deterge as a verb that means to wash off, or to cleanse. It defines detergent as a noun meaning cleansing substance. Therefore, there are many compounds that have these qualities and are referred to as soaps or detergents. Detergents are classed as surfactants. They may be anionic, cationic, and nonionic. The anionic surfactants (detergents) are the most popular of synthetic detergents because of their germicidal and cleansing effect. Thus, they are the greatest polluters. They ionize in water $(C_{11}H_{23}CO_2^-)$ to give an anion of large size and a cation (Na^+) of small size. In addition to the indicated classification, the chemist groups the synthetic detergents according to their synthesis processes and the primary structural compounds, which are branched compounds made by alkylating benzene with tetrapropylene (alkyl benzene sulfonate—ABS) and compounds made by alkylating benzene with *n*-paraffins (linear alkylate sulfonate—LAS). It has been shown that ABS compounds are only 40% to 70% biodegradable. As the result, they are known as biologically "hard" compounds. The LAS group have been shown to be up to 98% biodegradable; thus they are known as "soft" detergents. Many manufacturers add compounds known as "builders." These increase the cleansing effect of the synthetic detergents.

Of the large array, phosphate is one of the most popular and effective of the builders. An example is sodium tripolyphosphate. The builders and the paraffins are the major contributors to pollution because many treatment plants are unable to break them down chemically or cause them to recombine chemically into a harmless form. These side products cause foaming in water and in-terfere with the respiratory activities of the normal aquatic habital organism. The phosphates add to the nutrient content of the drainage area, thereby promoting the growth of bacteria and algae. These organisms may cause putrifaction and additional organic contamination of the drainage area or stream.

INSTRUCTOR: The reagents employed are those packaged by Hach Chemical Co., Ames, Iowa. Or to prepare the reagents, refer to *Standard Methods for the Examination of Water and Wastewater,* ed. 12, or ed. 13, p. 339, or *Analytical Chemistry* **38**:761, 1966.

Materials

1. Colorimeter or spectrophotometer
2. Sulfate buffer solution, detergent test powder pillows (crystal violet with carrier), toluene (reagent grade), methyl green powder pillows
3. Sample containers, 25, 100, and 500 ml graduate cylinders, 500 ml separatory funnel
4. Ring stands, clamps, circle clamps
5. Samples from streams, ponds, and places near sanitary treatment plant and/or one prepared by the instructor

Procedural information

The *crystal violet method* is relatively free from interference and only one extraction is required. In this method both types of detergents are determined: alkyl benzene sulfonate (ABS), the "hard" or nonbiodegradable type of detergent; and/or linear alkylate sulfonate (LAS), the "soft" or biodegradable type of detergent. Since ABS and LAS are generalized formulas, a difference may exist on a particular molecular weight form. *Acetone* is a suitable reagent for removing toluene from glassware. Perchlorate and periodate ions interfere.

Procedure

1. Measure 300 ml of the water sample in the 500 ml graduate cylinder; pour it into the *clean* 500 ml separatory funnel. Repeat, using 300 ml of *demineralized water* to be used for Spectronic 20 standardization.
2. Add 10 ml of the sulfate buffer solution to each. Stopper the separatory funnels and shake each for 5 to 8 seconds (shake gently).
3. Add the contents of one powder pillow of the detergent test powder to each 500 ml separatory funnel. Stopper and shake until the powder dissolves. For the DR colorimeter, use the contents of one methyl green powder pillow.
4. Add 30 ml of toluene to each separatory funnel and shake vigorously for 1 minute.
5. Put the separatory funnel on the ring stand and wait 15 minutes. The toluene will separate from the water and float. If detergents are present, the toluene layer will be blue.
6. After 15 minutes, drain off and discard the water of each sample.
7. Drain the toluene mixture into a clean

dry colorimeter bottle or 25 ml Erlenmeyer flask. Let stand for 20 minutes.

8. *DR colorimeter*—Fill a colorimeter bottle with the demineralized water sample and place it in the light cell. Insert the detergent meter scale and use the 4015 color filter, 615 nm. Adjust the light control for a meter reading of zero ppm.
9. *DR colorimeter*—Put the toluene sample into the light cell and read the ppm ABS or/and LAS detergents present. Post in the result table.
10. *B & L Spectronic 20*—After step 6, fill the ½-inch test tube two-thirds full of the sample.
11. *B & L Spectronic 20*—Standardize the instrument with the prepared demineralized water sample.
12. *B & L Spectronic 20*—Insert the ½-inch test tube that contains the sample, record the percent transmittances. Post in the result table.
13. *B & L Spectronic 20*—Determine the ppm anionic detergents from the B & L Spectronic 20 calibration table at the end of the exercise.
14. Refer to the Introduction for additional information on detergents.
15. Collect the information from the other students and post.

Result table

DR colorimeter

Samples	ppm detergents	Remarks
1		
2		
3		
4		

Result table

B & L Spectronic 20

Samples	Instrument reading (% T)	ppm detergents	Remarks
1			
2			
3			
4			

B & L SPECTRONIC 20 CALIBRATIONS
Anionic detergents (ppm)
615 nm, ½-inch test tube

	0	1	2	3	4	5	6	7	8	9
10	0.49	0.47	0.45	0.44	0.42	0.41	0.39	0.38	0.37	0.36
20	0.35	0.34	0.33	0.32	0.31	0.30	0.29	0.285	0.278	0.27
30	0.264	0.257	0.250	0.244	0.238	0.232	0.227	0.221	0.216	0.210
40	0.206	0.201	0.197	0.192	0.188	0.184	0.180	0.176	0.172	0.168
50	0.164	0.160	0.156	0.153	0.149	0.146	0.142	0.139	0.136	0.132
60	0.129	0.126	0.123	0.120	0.117	0.114	0.111	0.108	0.105	0.102
70	0.100	0.096	0.093	0.090	0.088	0.086	0.083	0.080	0.077	0.075
80	0.073	0.070	0.067	0.065	0.062	0.060	0.058	0.055	0.050	0.046
90	0.043	0.040	0.036	0.052	0.028	0.024	0.019	0.014	0.010	0.005

Questions

Account for the presence and variation of the quantity of detergents found to be present.

What effect do detergents have on the plants and animals that live in the water source?

Do they affect humans?

What chemical compounds make up the basic formulation of detergents?

How are detergents eliminated from our wastewater?

Name 10 detergents commonly found in the grocery store. Indicate the manufacturers' information about them.

16
Dissolved oxygen

INSTRUCTOR: There are several methods for the determination of dissolved oxygen (DO) in water. Refer to *Standard Methods for the Examination of Water and Wastewater,* ed. 12 or 13. However, this exercise uses the Hach Chemical Co. procedure and their prepared reagents.

Materals (The PAO method)

1. 25 ml beaker, glass-stoppered dissolved-oxygen bottle, 25 ml graduate cylinder
2. Hach No. 1079 PAO (phenylarsene oxide), dissolved oxygen powder I (manganous sulfate), dissolved oxygen powder II (alkaline iodide-azide), dissolved oxygen powder III (dry acid)
3. Four samples, including one from a sanitary facility

Procedural information

To avoid the possibility of air bubbles, incline the DO bottle slightly and insert the stopper with a quick thrust. The air bubbles will be forced out. *Do not allow the PAO solution to stand in direct sunlight.* If air bubbles are trapped in the DO bottle, discard and start over with another sample. After adding DO I and DO II and shaking, note that a flocculant precipitate will form. If O_2 is present, the precipitate will be brownish orange.

Procedure

1. Identify the samples according to number.

   1 _____
   2 _____
   3 _____
   4 _____

2. Slowly fill the dissolved oxygen bottle (DO bottle) with sample water. Be certain that *no air bubbles* are present.
3. Pour the contents of DO powder I pillow and the DO powder II pillow into the sample.
4. Stopper the DO bottle and shake. Watch for the precipitate to form.
5. Let the sample stand until the precipitate leaves the upper half of the bottle.
6. Shake and repeat step 5.
7. Remove the stopper and add the contents of DO powder III. Carefully re-stopper and shake. A yellow color will develop if O_2 is present.
8. Collect 6 ml of this prepared sample and pour it into the 25 ml Erlenmeyer beaker.
9. Using the PAO and cap dropper, add PAO a drop at a time while slowly swirling the sample. Count the number of drops until the sample turns from yellow to colorless.
10. Post in the result table the number of drops that were required for step 9. One drop is equal to 1 ppm of dissolved oxygen.
11. Collect the data for the other samples and post in the result table.
12. Refer to the text and Appendix for additional information on dissolved oxygen.

Result table

Sample	ppm dissolved oxygen	Remarks
1		
2		
3		
4		

Questions

How can you account for the differences in the ppm of dissolved oxygen of the samples?

What are the recommended levels of dissolved oxygen for each source of the sample?

Why do the recommendation levels vary? This should indicate the need for oxygen in the water source.

Are the various levels of dissolved oxygen necessary in our water sources to maintain a high standard of environmental water quality? Explain.

What conditions may prevail in the sample sources that have a low quantity of oxygen?

PART III

Microbiological examination of water

Natural water and wastewater contain a variety of different types of microorganisms. Water can serve as a carrier for thousands of different types. Bacteria play a significant role in the potability of the water that is available for human consumption. They are also used as indicators of the degree of pollution that exists in the lakes, rivers, streams, or ground water.

The greatest population of this diversity of organisms is represented by the saprophytic and parasitic groups. Within each of these types, we find species that are pathogenic to plants, animals, and man. The Enterobacteriaceae family of bacteria is of major concern to man and his environment. This family includes the enteric genera, some of which are considered pathogens, such as *Salmonella* and *Shigella,* as well as the nonpathogenic coliforms, such as *Escherichia coli* and *Aerobacter aerogenes.*

The saprophytic group of bacteria are those that are capable of biodegrading complex organic matter into compounds that provide the nutritional requirements for their own metabolic processes. In addition to supplying nutrients for themselves, their metabolic activities contribute to the converting of complex organic matter into forms that can be assimilated by plants, which provide a food source for animals. Thus, there is an "integrated cycle of life" between bacteria, plants, and animals. It would be impossible for continued life on this planet if the saprophytic bacteria were not present. The presence of a large population of saprophytic bacteria furnishes a guide to the microbiological quality and palatability of the water. The density of this group of bacteria is directly proportional to the quantity of organic matter and dissolved solids in the lakes, rivers, streams, and ground water. Therefore, the importance of this group is twofold: (1) their population density serves as a pollution indicator, (2) there are genera, such as *Salmonella, Shigella, Bacillus,* and *Clostridium* that are pathogenic to man and animals.

The parasitic group of bacteria are unable to synthesize all of their nutritional requirments. They must live within a host that supplies them with the necessary nutrients for their metabolic activities. In some cases, the host is dependent on the by-products of the parasite. In other associations, the host may not assimilate the food into its metabolic cycles until the parasites have decomposed the food into a very simple molecular form. In many mutualisms, both the host and parasite benefit to an equal degree. The parasites of the enteric group that affect man are those that are used to determine our water quality. The presence of these bacteria indicate that the water has been contaminated (polluted) by human or animal fecal wastes. The enteric bacteria are not able to reproduce and live in water for very long periods. However, they may survive for several days, thus providing an opportunity for them to be consumed by humans. Once within their normal habitat, they will propagate rapidly. Thus, if an infectious species is present, a disease condition or epidemic may develop. Some of the common diseases that are spread by the presence of the enteric group that may be found in water are the intestinal infections: Asiatic cholera, amebic dysentery, bacillary dysentery, and typhoid fever.

The coliform group (enteric bacteria) represents a large heterogeneous group that normally inhabit the intestinal tract of warm-blooded animals. They may be highly pathogenic, slightly pathogenic, or nonpathogenic. The *Standard Methods for the Examination of Water and Wastewater* (ed. 12 or 13) defines the coliform group as follows: "The coliform group includes all of the aerobic and facultative anaerobic, Gram-negative, nonspore-forming, rod-shaped bacteria which ferment lactose with gas formation within 48 hours at 35° C." Based upon this definition, it is obvious that many different genera are included. Some of the common coliforms include *Escherichia coli, E. aurescens, E. freundii, E. intermedia,* and *Aerobacter cloacae. Klebsiella* are organisms that have coliform characteristics. They normally inhabit the soil and water instead of the intestinal tract of man. However, they are usually found to be present when testing for *E. coli.* Some experts in the field consider *Klebsiella* as a potentially dangerous pathogen.

The coliform bacteria are used as pollution-indicator organisms. If the food or water sources have been contaminated with animal feces, a large population of coliforms will be present, such as 2 to 5 million bacteria per 100 ml of water sample. This is a direct indication that the fecal matter may harbor pathogenic bacteria that have been discharged with the feces of humans affected with a bacterial disease.

The standard bacterial plate determines the total population of bacteria. The presence of a large number of bacteria indicates the degree of pollution of the water by plant and animal tissue, fecal matter, and other organic material.

The coliform tests provide information that indicate the degree of pollution by animal excreta. It also serves as an indicator for the possible presence of disease-causing organisms. The desirable population of coliform bacteria in water used for human consumption is 100 bacteria per 100 ml of water. Refer to Surface Water Criteria for Public Water Supplies, found in the Appendix, p. 132.

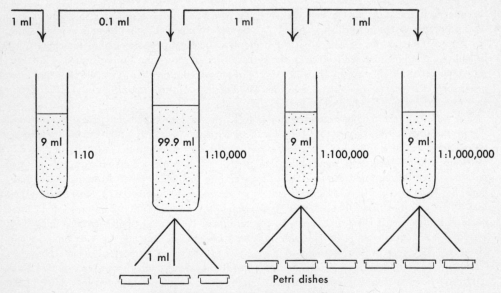

Fig. 6. Diagram of dilution procedures.

17
Standard bacterial plate count

INSTRUCTOR: The quantity of materials is determined by the number of students performing the exercise. Refer to the *Difco Manual* or any microbiology laboratory manual for the preparation of the media.

Materials

1. Autoclave, incubator, two 500 ml graduate cylinders, 90 mm × 15 mm sterile disposable plastic petri dishes, 1 ml delivery 0.1 gradations and 10 ml disposable plastic or glass pipettes, 125 ml Erlenmeyer flasks
2. Sterile dilution blanks that contain 9 ml and 99.9 ml of sterile distilled water (see the preceding page and below for dilution procedures)
3. 1000 ml prepared tryptone glucose yeast agar or nutrient agar
4. Collect four samples and include a sample from the sanitary facility

Procedure

1. Identify samples according to number.

1 _____. 3 _____.

2 _____. 4 _____.

2. Make the following dilutions: 1:10; 1:10,000; 1:100,000 (see Fig. 6).
3. Liquefy the prepared agar by immersing the flask in boiling water.
4. Cool the agar in a water bath to 35° to 45° C.
5. Swirl your sample to get even distribution of the bacteria.
6. With the sterile 1 ml pipette, transfer 1 ml of the sample to a petri dish.
7. Add about 10 ml of the medium that has been cooled to 35° to 45° C.
8. Slowly swirl the material to get an even distribution.
9. After the medium has *solidified*, place the petri dish in an inverted position into the incubator set at 35° C. Incubate for 24 to 48 hours.
10. Determine the population of bacteria per 1 ml of sample. If a dilution is used, such as 1:10,000, multiply the number of bacterial colonies by the dilution factor to determine the number of bacteria per milliliter.

WATER TESTING DILUTIONS (Fig. 6)
Procedure

1. Take 1 ml of sample and add 9 ml of sterile distilled water. Shake vigorously for 5 seconds. (Dilution 1:10 or 10%.)
2. Take 0.1 ml of 1:10 dilution and add to 99.9 ml of sterile distilled water and then shake.* (Dilution 1:10,000 or 0.01%.)
3. Take 1 ml of 1:10,000 dilution and add to 9 ml of sterile distilled water and then shake.* (Dilution 1:100,000 or 0.001%.)
4. Take 1 ml of 1:100,000 dilution and add to 9 ml of sterile distilled water and then shake.* (Dilution 1:1,000,000 or 0.0001%.)

*Plate 1 ml in triplicate series of each dilution.

Result table
Representing the number of bacteria per
milliliter of sample

| Samples | Dilution and populations | | | Remarks |
	1:10	1:10,000	1:100,000	
1				
2				
3				
4				

Questions

Why were the dilutions made?

Compare the different shapes and colors of the colonies found in each sample. What are the causes for these differences?

Is there evidence of growth of organisms other than bacteria? Give examples.

Will all the bacteria present in the samples grow in this medium? Explain your answer.

What is the significance of the population of bacteria in the source of each sample?

What are the possible sources of nutrients that account for the presence of a large population of bacteria?

Does this experience reflect an important pollution parameter?

18
Differential test for the coliform group of bacteria

Eosin methylene blue agar plate count is the method of value for the separation of members of the *Escherichia* and *Aerobacter* divisions of the colon group.

INSTRUCTOR: Refer to the *Difco Manual* or any microbiology laboratory manual for the preparation of the medium. The number of students performing the test will determine the quantity of material and equipment necessary.

Materials

1. The *same equipment necessary for the standard plate count*
2. Dilutions of the samples used in the standard plate count
3. Eosin methylene blue agar medium

Procedure

1. Follow the same procedure steps 1 to 9, as outlined for the standard bacterial plate count.

2. After 24 hours of incubation, examine the culture and count the number of colonies that possess a brilliant metallic sheen.
3. If no colonies are present, reincubate for an additional 24 hours.
4. Use step 10 of the standard bacterial plate count method to determine the total population.
5. Count the colonies of *Escherichia coli* that show dark centers and a greenish metallic sheen.
6. Count the colonies of *Aerobacter aerogenes* that show brown centers but may not possess a metallic sheen.
7. Transfer the results of any dilution of the standard bacterial plate count to the table of this exercise.
8. Post the numerical results for steps 4, 5, and 6 in the result table.

Result table
Representing the number of bacteria per milliliter of sample

Samples	* †(g)SBPC 1	Dilutions and populations			
		1:10	1:10,000	1:100,000	% Coliform
1					
2					
3					
4					

*Dilution from standard bacterial plate count.
†(g) refers to the population of the standard bacterial plate count of an indicated dilution.

Questions

Is there evidence of growth by organisms other than bacteria? What are they?

Why is it necessary to distinguish between *E. coli* and *A. aerogenes?*

Why must the cultures be incubated at about 35° C.?

What significance does a high population of coliform bacteria have on environmental quality?

Make as many deductions as possible by studying the table and comparing the population results of both tests.

Are pathogenic fecal bacteria able to grow in this medium?

Describe the testing methods and procedures that are recommended by the *Standard Methods for the Examination of Water and Wastewater* (ed. 12, 1965, or 13, New York, 1971, American Public Health Association).

Explain the value of this test for determining the presence of pathogenic bacteria.

19
Most probable number of coliform bacteria (presumptive and confirmative tests)

INSTRUCTOR: There are numerous methods and procedures available to determine coliform bacterial populations. Most of the procedures have been adopted from *Standard Methods for the Examination of Water and Wastewater,* ed. 12 or 13. If it is desirable to assemble the material and necessary reagents, refer to Part VII of *Standard Methods.* The procedure for media preparation can be found in the *Difco Manual* or any microbiology laboratory manual.

This exercise follows the Hach Chemical Company procedure. This procedure is highly recommended for coliform determination by municipalities and industry. It eliminates the necessity for apparatus assembly, media, and reagent proportions. *Discuss this exercise with the students prior to performing it in class.*

Materials

1. Incubator, test tube holders
2. Ten each, disposable sterile pipettes, 1 ml, 10 ml, and 20 ml
3. Five, 99 ml sterile water blanks, sterile funnels
4. 25 Coliform presumptive tubes with lactose broth (dehydrated)
5. 25 Coliform confirmative tubes with brilliant green lactose bile broth

GENERAL DIRECTIONS FOR THE PRESUMPTIVE AND CONFIRMATIVE TUBES
Presumptive test tubes

1. Each student or group collects a total of four samples and identifies each of them accordingly.

    1 _____. 2 _____. 3 _____. 4 _____.

2. Label the water sample and dilution samples. Use four dilutions plus the original sample.
3. Wash hands thoroughly with soap in order to minimize contamination of the equipment.
4. Cut the sealed package and remove the presumptive coliform test assembly.
5. Remove the cap while being careful not to touch the inside of the cap or the open end of the tube assembly.
6. Fill the tube with the water sample or *dilutions of the sample* (use a sterile funnel or sterile 20 ml volumetric pipette).
7. Replace the screw cap and allow a few minutes for the dehydrated nutrient medium to dissolve (periodic shaking of the tube will hasten solution). Then invert the tube assembly to allow the smaller inner glass vial to fill with water. Hold the tube assembly upright and check to see that all air bubbles are removed from the small inverted vial.
8. Place the tube in the proper row of the holder. After 1 hour, examine the inverted vial in the tube for entrapped air. Some water samples contain high concentrations of dissolved air, and small air bubbles will develop as the temperature of the water is raised. If there are air bubbles, invert the tube assembly briefly to allow the bubbles to escape from the inverted vial. Return the tube assembly to the incubator. Thereafter, the tube assembly must be kept in an upright position at all times.
9. Examine the tube after 24 ± 2 hours. If a gas bubble has collected in the in-

verted vial, coliform bacteria are presumed to be present. If no gas has collected, allow the sample to remain in the incubator and examine it again after a total time 48 ± 2 hours has elapsed. Formation of gas at either time in any amount in the small inside vial constitutes a positive presumptive test. The absence of gas formation at the end of this period constitutes a negative test for this tube assembly.

Confirmative test tubes

There are a few unusual types of bacteria outside of the coliform group that will produce a positive test with the *lactose broth tubes*. Therefore, it is necessary to perform a confirmative test to verify the presence of coliform bacteria. This test is performed with the use of a tube assembly filled with *brilliant green lactose bile broth*. The procedure is described as follows:

1. Wash hands thoroughly with soap in order to minimize contamination of the equipment.
2. Cut the sealed package and remove the confirmative test assembly.
3. When gas formation is first observed in a presumptive unit, shake or invert the tube in order to wet the rubber liner of the cap. Exchange this cap with the cap of a new Confirmation ColiVer Tube Assembly, brilliant green lactose bile broth. This serves to transfer some of the bacteria to the confirmative tube. Invert the confirmative tube to complete the transfer of bacteria from the cap into the liquid. *This step should be repeated for each positive presumptive tube.*
4. Inspect each confirmative unit to be certain no air is initially entrapped in the inverted inner vial.
5. Return to the same position that the coliform presumptive tube occupied.
6. Maintain the confirmative tubes at 95° F for 48 ± 2 hours. A positive confirmative test for coliform bacteria is indicated by gas collected in the inverted vials within the assembly units. No evidence of gas indicates a negative test.

Procedure for presumptive and confirmative tests
DILUTION SERIES

1. 10 ml sample
   No dilution. Shake the sample vigorously 25 times and then fill five presumptive ColiVer units, containing the lactose broth, to the top with the sample. Cap each and invert each unit to fill the inner tube. Place these five tubes in row A positions 1 to 5 of the holder.

2. 1 ml sample or 10^{-1} dilution
   A 1:10 dilution is conducted by adding 11 ml of the sample to 99 ml of sterile dilution water, by using a sterilized 11 ml pipet. Shake vigorously 25 times. Label it 10^{-1} dilution. Fill another five presumptive ColiVer units, containing the lactose broth, with this dilution. Cap and invert to fill the inner tube. Add this series to row B of the older.

3. 0.1 ml (10^{-1} ml sample or 10^{-2} dilution)
   A 1:100 dilution is conducted by adding *1 ml* of the sample to a 99 ml sterile dilution water by a sterile 1 ml pipette. Shake vigorously 25 times. Label 10^{-2} dilution. Fill five presumptive ColiVer units, containing the lactose broth, with this dilution. Cap and invert to fill inner tubes. Place this dilution in row C of the holder.

4. 0.01 ml (10^{-2} ml sample or 10^{-3} dilution)
   Take 1 ml of the 10^{-1} dilution from step 2 of the procedure, add to 99 ml of sterile water. Shake vigorously 25 times and label 10^{-3} dilution in (10^{-1} + 10^{-2}) of 10^{-3}. Fill another five presumptive ColiVer units, containing the lactose broth, with this dilution. Cap and invert to fill the inner tube. Place this dilution in row D of the holder.

5. 0.001 ml (10^{-3} ml sample or 10^{-4} dilution)
   Take 1 ml of the 10^{-1} sample (10^{-2} dilution), using a sterile 1 ml pipette. Add to a 99 ml of sterile dilution water. Shake vigorously 25 times and label 10^{-4} dilution. Fill another five presumptive ColiVer units, containing the lactose broth, with this dilution. Cap and invert to fill the inner tubes. Place this dilution in row E of the holder.

If other dilutions are necessary, the initial dilutions would not require setting up of ColiVer units, since these will or are assumed to be all positive.

The general procedure involves 1 ml + 99 ml dilution of either the sample or previous dilutions. Remember to add exponents of each dilution, that is, if 1 ml of the 10^{-6} dilution is diluted to 99 ml, this dilution is 10^{-8} ($10^{-6} + 10^{-2}$). The number of milliliters of sample is equal to one power of ten less than the dilution—in this example, therefore, 10^{-7} ml of sample.

At some point in the series, either at the beginning for less polluted waters or next to the last for polluted water, a 1:10 dilution is required. For this dilution, 11 ml of the sample or previous dilution are added to 99 ml. So, if the previous dilution which was used was the 10^{-6} dilution, this dilution is 10^{-7} ($10^{-6} + 10^{-1}$), and as above the volume of sample is 10^{-6} ml.

At the conclusion of this part we have 3-5 dilutions in the ColiVer presumptive tubes of the sample ready for incubation. GOOD RECORD KEEPING IS ESSENTIAL. Indicate the volumes to be incubated. For example, see Data Sheet 2, p. 69.

1. Place the holder in a 35° C incubator. Note time. After about 1 hour, examine the inverted vial in the tube for entrapped air caused by dissolved oxygen released in heating. If air bubbles are observed, holding hand on top of all tubes, invert the holder, or if only a few tubes show dissolved air, these tubes may be individually inverted and returned to their positions.
2. After 24 ± 2 hours, check each sample. The collection of gas in the inner tube is indicative of a positive presumptive test for the coliform bacteria. To confirm its presence, invert the presumptive tube to wet the cap, remove the cap of both the presumptive and a confirmed ColiVer tube (BGLBB), and exchange caps to inoculate the confirmative tube. CAUTION: *Do not touch the lips of the tubes.* Invert the confirmative tube. Inversion fills the inner tube and inoculates this tube. Return to the *same* position as the presumptive tube. Discard the *positive* presumptive tubes. Record as positive as shown in Data Sheet 2. After all presumptive tubes that were positive are replaced by inoculated confirmative tubes, return the holder to the incubator. It now holds both confirmative and presumptive tubes.
3. After another 24 hours, remove the holder from the incubator. First check the presumptive tubes. Follow the same procedure of inoculation of the confirmative tubes as on the first day. Record as negative all presumptive tubes that do not show gas formation. These tubes may now be discarded. Now check the confirmative tubes. Record as positive all tubes that show gas formation. These tubes may be discarded. Return to the incubator the confirmative tubes that have shown no gas formation.
4. After another 24 ± 2 hours, check the remaining confirmative coliform tubes for gas formation. Record as positive all tubes showing gas formation. The confirmative tubes that were set up after the first day are complete. Record as negative and discard.
5. The confirmative tubes that were set up after the second day and that have not shown gas formation must be incubated for another period of 24 ± 2 hours. At the end of this period, record as positive or negative and discard all tubes.

USE OF MPN (MOST PROBABLE NUMBER) TABLES

Using probability mathematics, one can estimate the number of coliform organisms producing any combination of positive and negative results. The MPN table is based on *3 sample volumes of 5 replicates* in decreasing decimal increments, such as 10 ml, 1 ml, 10^{-1} ml.

Coding

If five 10 ml portions, five 1 ml portions, and five 0.1 ml portions are inoculated initially, and positive results are secured from *five* of the *10 ml portions, three* of the *1 ml*

DATA SHEET I

Collection data

Sample description_____

Sample number_____

Data collected_____

Time collected_____

Collected by_____

Laboratory data

Date_____

Time_____

By_____

Row	Number	ml sample	Presumptive 24 hours	Presumptive 48 hours	Confirmed 24 hours	Confirmed 48 hours	Remarks
A	1						
	2						
	3						
	4						
	5						
B	1						
	2						
	3						
	4						
	5						
C	1						
	2						
	3						
	4						
	5						
D	1						
	2						
	3						
	4						
	5						
E	1						
	2						
	3						
	4						
	5						

Confirmation code_____ (MPN/100 ml) [MPN = most probable number]

Coliform_____

Data sheet 2

Collection data

Sample description _____

Sample number _____

Data collected _____

Time collected _____

Collected by _____

Laboratory data

Date _____

Time _____

By _____

Row	Number	ml Sample	Presumptive 24 hours	Presumptive 48 hours	Confirmed 24 hours	Confirmed 48 hours	Remarks
A	1		+		+		
	2		+		+		
	3		+		+		
	4		+		+		5
	5		+		+		
B	1	1	+		−	+	
	2		+		+		
	3		−	+	+		5
	4		+		+		
	5		+		+		
C	1	10^{-1}	−	−			
	2		−	+	+		
	3	0.1	+		+		3
	4		−	+	−	+	
	5	1:10	−	−			
D	1	10^{-2}		+	+		
	2	0.01	−	−			
	3		−	−			2
	4	1:100	+		+		
	5		−	−			
E	1	10^{-3}	−	−			
	2	0.100	−	−			
	3		−	−			0
	4	1:1000	−	−			
	5		−	−			

Confirmation code   5, 3, 2   (1)

Coliform   1410

(MPN/100 ml)

$$141 \times \frac{10}{1} = 1410$$

portions, and *none* of the *0.1 ml portions,* then the *coded result* of the test is *5-3-0.* The code can be looked up in the *MPN Table,* and the MPN per 100 ml is recorded directly. If more than the above three sample volumes are to be considered, then the determination of the coded result may be more complex. The *examples* described in Table 1 are useful *guides* for selection of the significant series of three sample volumes.

Discussion of examples (refer to last column in Table 1)

1. When all the inoculated tubes of more than one of the decimal series give positive results, then it is customary to *select the smallest sample volume (here, 10 ml) in which all tubes gave positive results.* The results of this volume, and the next lesser volumes are used to determine the coded result.

2. When none of the sample volumes gives positive results in all increments of the series, then the results obtained are used to designate the code. Note that it is *not* permissible to assume that if the next larger increment had been inoculated, all tubes probably would have given positive results, and therefore assign a 5-4-1 code to the results.

3. Here the results are spread through four of the sample volumes. In such cases, the number of positive tubes in the smallest sample volume is added to the number of tubes in the third sample volume (counting down from the smallest sample volume in which all tubes gave positive results).

4. Here it is necessary to use the 5-5-4 code, because inoculations were not made of 0.001 ml sample volumes; and it is *not* permissible to assume that if such sample volumes had been inoculated, they would have given negative results, or any other arbitrarily-designated result.

5. This is an indeterminate result. Many MPN tables do not give a value for such a result. If the table used does not have the code, then look up the result for code 5-5-4, and report the result "greater than" the value shown for the 5-5-4 code.

6. Like discussion number 5, this is an indeterminate result. If the code does not appear in the table being used, then look up the result for code 1-0-0, and report the MPN as "less than" the value shown for the 1-0-0 code.

7. The twelfth and thirteenth editions of *Standard Methods* stipulate this type of code designation when unusual results such as this occur.

Table I. Examples of coded results

Number of ml sample per tube	100	10	1	0.1	0.01	0.001		See discussion
Number of tubes per sample volume	5	5	5	5	5	5	Code	number
Number of tubes in sample giving		5	4	1			5-4-1	
positive results in test	5	5	4	0	0	0	5-4-0	(1)
		4	1	0	0	0	4-1-0	(2)
	5	5	4	1	1	0	5-4-2	(3)
		5	5	5	4		5-5-4	(4)
		5	5	5	5		5-5-5	(5)
		0	0	0	0		0-0-0	(6)
		0	1	0	0		0-1-0	(7)
		1	0	0	0		1-0-0	(8)

Table 2. MPN (most probable number) and 95% confidence limits for various combinations of positive results in a planting series of five 10-ml, five 1-ml, and five 0.1-ml portions of sample

Number of tubes giving positive reaction out of			MPN index (organisms per 100 ml)	Confidence limits (95%)	
Five 10-ml portions	Five 1-ml portions	Five 0.1-ml portions		Lower limit	Upper limit
0	0	0	<2		
0	0	1	2	<0.5	7
*0	0	2	4	<0.5	11
0	1	0	2	<0.5	7
*0	1	1	4	<0.5	11
*0	1	2	6	<0.5	15
0	2	0	4	<0.5	11
*0	2	1	6	<0.5	15
*0	3	0	6	<0.5	15
1	0	0	2	<0.5	7
1	0	1	4	<0.5	11
*1	0	2	6	<0.5	15
*1	0	3	8	1	19
1	1	0	4	<0.5	11
1	1	1	6	<0.5	15
*1	1	2	8	1	19
1	2	0	6	<0.5	15
*1	2	1	8	1	19
*1	2	2	10	2	23
*1	3	0	8	1	19
*1	3	1	10	2	23
*1	4	0	11	2	25
2	0	0	5	<0.5	13
2	0	1	7	1	17
*2	0	2	9	2	21
*2	0	3	12	3	28
2	1	0	7	1	17
2	1	1	9	2	21
*2	1	2	12	3	28
2	2	0	9	2	21
*2	2	1	12	3	28
*2	2	2	14	4	34
2	3	0	12	3	28
*2	3	1	14	4	34
*2	4	0	15	4	37

*Omitted from the twelfth and thirteenth editions of *Standard Methods for the Examination of Water and Wastewater.*

Continued.

Table 2. MPN (most probable number) and 95% confidence limits for various combinations of positive results in a planting series of five 10-ml, five 1-ml, and five 0.1-ml portions of sample—cont'd

Number of tubes giving positive reaction out of			MPN index (organisms per 100 ml)	Confidence limits (95%)	
Five 10-ml portions	Five 1-ml portions	Five 0.1-ml portions		Lower limit	Upper limit
3	0	0	8	1	19
3	0	1	11	2	25
*3	0	2	13	3	31
3	1	0	11	2	25
3	1	1	14	4	34
*3	1	2	17	5	46
*3	1	3	20	6	60
3	2	0	14	4	34
3	2	1	17	5	46
*3	2	2	20	6	60
3	3	0	17	5	46
*3	3	1	21	7	63
*3	4	0	21	7	63
*3	4	1	24	8	72
*3	5	0	25	8	75
4	0	0	13	3	31
4	0	1	17	5	46
*4	0	2	21	7	63
*4	0	3	25	8	75
4	1	0	17	5	46
4	1	1	21	7	63
4	1	2	26	9	78
4	2	0	22	7	67
4	2	1	26	9	78
*4	2	2	32	11	91
4	3	0	27	9	80
4	3	1	33	11	93
*4	3	2	39	13	106
4	4	0	34	12	96
*4	4	1	40	14	108
*4	5	0	41	14	110
*4	5	1	48	16	124
5	0	0	23	7	70
5	0	1	31	11	89
5	0	2	43	15	114
*5	0	3	58	19	144
*5	0	4	76	24	180
5	1	0	33	11	93
5	1	1	46	16	120
5	1	2	63	21	154
*5	1	3	84	26	197

Table 2. MPN (most probable number) and 95% confidence limits for various combinations of positive results in a planting series of five 10-ml, five 1-ml, and five 0.1-ml portions of sample—cont'd

Number of tubes giving positive reaction out of			MPN index (organisms per 100 ml)	Confidence limits (95%)	
Five 10-ml portions	Five 1-ml portions	Five 0.1-ml portions		Lower limit	Upper limit
5	2	0	49	17	126
5	2	1	70	23	168
5	2	2	94	28	219
*5	2	3	120	33	284
*5	2	4	148	38	366
*5	2	5	177	44	515
5	3	0	79	25	187
5	3	1	109	31	253
5	3	2	141	37	343
5	3	3	175	44	503
*5	3	4	212	53	669
*5	3	5	253	77	788
5	4	0	130	35	302
5	4	1	172	43	486
5	4	2	221	57	698
5	4	3	278	90	849
5	4	4	345	117	999
*5	4	5	426	145	1161
5	5	0	240	68	754
5	5	1	348	118	1005
5	5	2	542	180	1405
5	5	3	920	210	3000
5	5	4	1600	350	5300
*5	5	5	2400	800	—

8. Note the difference from discussion number 7 above. Inoculations of 100 ml portions were not made, and it cannot be assumed that the result would have called for code 0-1-0.

Computing and recording the MPN (most probable number)

When the dilution tube results have been codified, they are read and recorded from Table 2. Post results in the result table.

1. If, as in the first discussion of examples shown on p. 70, the first number in the coded result represents a 10 ml sample volume, then the MPN per 100 ml is read and recorded directly from the appropriate column in the table.

2. On the other hand, if the first number in the coded result represents a sample volume other than 10 ml, then a calculation is required to give the corrected MPN. For example in the fourth example on p. 70, the 5 of the 5-5-4 code represents a sample volume of 1 ml. Look up the 5-5-4 code *as if* the 1 ml volume actually were 10 ml, *as if* the 0.1 ml volume actually were 1 ml, and *as if* the 0.01 ml volume actually were 0.1 ml. The MPN obtained (1600) then is multiplied by a

factor of 10 to give the corrected value. A simple formula for this type of correction is:

$$\underset{\substack{\text{MPN} \\ \text{(from} \\ \text{table)}}}{} \times \frac{10}{\text{Largest ml sample tested}} = \text{MPN/100 ml}$$

Example: From a sample of water, 5 out of five 0.01 ml (10^{-2} ml) portions, 2 out of five 0.001 ml (10^{-3} ml) portions, and 0 out of five 0.0001 ml (10^{-4} ml) portions gave positive reactions.

From the code 5-2-0 in the MPN table, the MPN index is 49

$$\underset{\substack{49 \\ \text{(from table)}}}{} \times \frac{10}{0.01} = 49,000$$

$$\text{MPN index} = 49,000$$

Result table

Samples	MPN	Federal standard	Difference	Remarks
1				
2				
3				
4				

Questions

Identify the possible sources of *Escherichia coli* (coliform) for each sample.

How can the population of coliforms be controlled?

What is the effect of high populations of coliform bacteria in each of the water sources? On our environment?

Adapted from Hach Chemical Company procedure for most probable number coliform bacteria. Portions of this procedure were taken from *Current Practices in Water Microbiology, Training Course Manual,* January 1968, Federal Water Pollution Control Administration. Outlines 20 and 22 prepared by H. L. Jeter, Chief FWPCA Training Activities, SEC.

20
Oxygen demand index

The Illinois Department of Public Health developed the oxygen demand index as a quick simplified analysis of water to give an estimated normal 5-day biochemical oxygen demand (BOD). (For an explanation of the significance of BOD, refer to the introduction to the manometric BOD Test.) The process is based upon the BOD value of glucose. This value is used for calibration (caution must be used to perform this experiment). Refer to *The Digester* **22**:4, 1965, Illinois Department of Public Health for additional information.

Materials

1. Colorimeter or spectrophotometer (600 nm); if spectrophotometer is used, prepare the samples in a 25 ml Erlenmeyer flask and use colorimeter filter No. 3482
2. Potassium dichromate in sulfuric acid, mercuric sulfate, concentrated sulfuric acid with catalytic amounts of silver sulfate
3. Samples from a sewage treatment plant, lake, stream, or pond
4. Water bath, hot plate, 25 and 100 ml graduate cylinders

Procedure

1. Into a colorimeter bottle put 10 ml of demineralized water.
2. Put 10 ml of the sample into another colorimeter bottle.
3. To each of the colorimeter bottles prepared in steps 1 and 2, add 4 ml of potassium dichromate–sulfuric acid solution. (These are dangerous chemicals; use caution. Do not mouth pipette.)
4. To each of the colorimeter bottles, add 0.1 gram of mercuric sulfate and mix; use a measured scoop for this addition.
5. Next add 14 ml of sulfuric acid–silver sulfate solution. (CAUTION: The bottle will get very hot. Swirl to mix.)
6. Using a *rack,* place the bottle in a boiling water bath for 20 minutes and maintain the bath at boiling temperature, 100° C. (CAUTION: The colorimeter bottles are not Pyrex; therefore, they will crack if removed from the bath and put on a room temperature surface.) Remove the heat from the water bath and allow to stand until the water reaches room temperature.
7. To standardize the DR colorimeter, insert the *blank* bottle into the light cell, using the appropriate ODI meter scale and recommended filter, and adjust the light control for a meter reading of zero ppm.
8. Insert the prepared sample into the light cell and read ppm ODI and post in the result table.
9. To standardize the B & L Spectronic 20, use the prepared demineralized water sample. Read the % T. Find the ODI from the following calibration table and post in the result table.

B & L SPECTRONIC 20 CALIBRATIONS
600 nm, ½-inch test tube

	0	1	2	3	4	5	6	7	8	9
60	450	430	417	412	391	379	365	350	339	328
70	312	300	291	279	265	258	242	232	221	210
80	195	185	175	165	155	140	130	120	110	100
90	90	80	70	60	50	40	33	15	10	5

Result table

Samples   ppm ODI

1 _____

2 _____

3 _____

4 _____

Questions

Define and explain the significance of ODI.

How do you account for the differences in values?

What do you believe are the contributors to the estimated BOD results?

What is the significance of this test in determining environmental quality as it may affect or water supply? Explain fully.

76

21
Biochemical oxidation demand

Environmental water quality depends on the removal of the biochemical oxygen demand (BOD) from the water supply and wastewaters discharged into streams, lakes, and rivers. Therefore, the quality of the water depends on the removal of BOD from sewage, industrial wastes, storm water and drinking water sources. It is imperative that the degree of pollution by the constituents that determine the BOD be constantly measured and monitored. This is accomplished by chemical methods for BOD analysis or by a manometric BOD apparatus analysis.

INSTRUCTOR: The principal chemical method of BOD analysis depends on the measurement and dilution of a water sample into a series of separate bottles that contain a known concentration of dissolved oxygen. These bottles are then incubated at a constant temperature of 20° C. for 5 days. After this period, the residual dissolved oxygen is measured. The loss of dissolved oxygen during the incubation period, multiplied by the dilution factor, gives the 5-day BOD of the sample tested. Complete details for the proper procedure are given in the *Standard Methods for the Examination of Water and Wastewater* (ed. 13, New York, 1971, American Public Health Association). The described procedures must be performed by trained technicians because of the extreme degree of accuracy needed in the preparation and handling of the corrosive and toxic chemicals used in the preparation of the various reagents.

The adoption of the Hach Manometric BOD for use in an introductory environmental science course provides a method for performing the BOD test that has several advantages over the *Standard Methods* procedure. These advantages are that no special training in chemistry is needed, it is not necessary to prepare reagents analytically, in most cases multiple dilutions are not necessary, chemical measurement of DO is unnecessary, the results are read directly in terms of the BOD concentration without the need for calculation of dilutions, and preconditioning of samples is not normally needed.

Refer to the *Hach Manometric BOD Apparatus Instruction Manual* before introducing this exercise to the students. It is suggested that the instructor set up the assembly and determine the pretesting sample treatments, if needed, as outlined in the manual. For best classroom results and student interpretation of the significance of BOD, use the procedure as outlined in this exercise.

Procedural information

This test is for unchlorinated samples requiring no preconditioning such as domestic waste, river water, streams, ponds, and sewage effluent. Refer to *Standard Methods* if the samples are chlorinated and for other factors that may affect the BOD results.

When a measured sample of the test material containing bacteria is placed in the bottles, as outlined in the procedure, and they are connected to a closed-end mercury manometer, a biological oxidation system is produced. The oxygen consumed by the bacterial metabolic processes can be accurately measured over time. As the oxygen is consumed by the biological processes, a pressure drop is indicated on the manometer. By supplying the system with a known volume of sample, as indicated in Table 3, the oxygen consumed

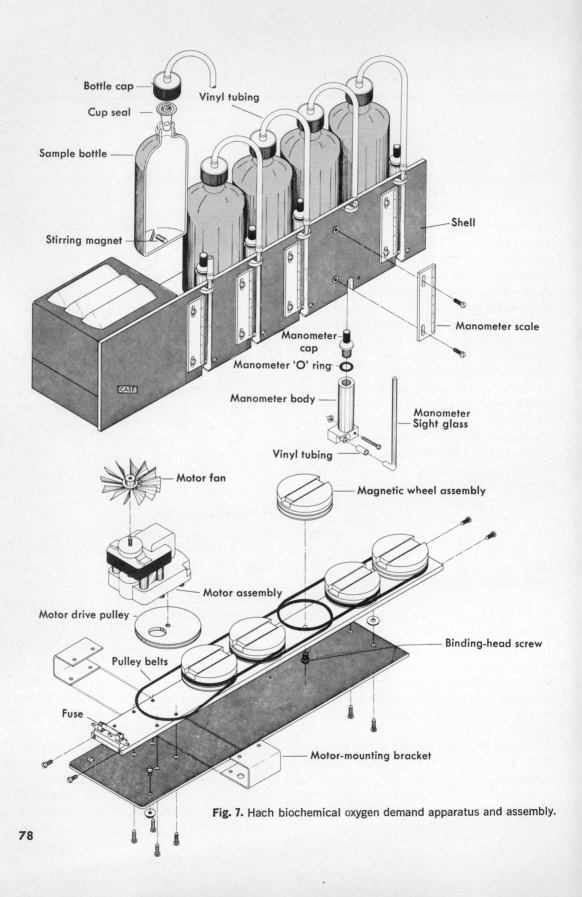

Bottle cap

Vinyl tubing

Cup seal

Sample bottle

Shell

Stirring magnet

Manometer cap

Manometer scale

Manometer 'O' ring

Manometer body

Manometer Sight glass

Vinyl tubing

Motor fan

Magnetic wheel assembly

Motor assembly

Motor drive pulley

Binding-head screw

Pulley belts

Fuse

Motor-mounting bracket

CASE

Fig. 7. Hach biochemical oxygen demand apparatus and assembly.

Table 3

Expected BOD range	Sample volume needed (milliliters)	Multiply result by
0– 35	428	0.1
0– 70	360	0.2
0– 175	244	0.5
0– 350	157	1
0– 700	94	2
0–1400	56	4
0–3500	21.7	10

by the bacteria is reflected in a pressure drop that is related directly as BOD.

The following BOD determination procedure will provide a 5-day, 20° C BOD of the sample sources.

Materials

1. Hach Manometric BOD Apparatus (see Fig. 7), incubator or refrigerator set at 20° C
2. Potassium hydroxide, 45% solution
3. Samples from five different sources

Procedure

1. Identify the samples according to number.

   1 _____. 4 _____.

   2 _____. 5 _____.

   3 _____.

2. Measure carefully the correct volume (usually 157 ml) of the well-mixed sample into one of the cleaned brown bottles. (See Table 3 for other recommended volumes.) Sample temperature should be within 2° C of the subsequent incubation temperature. Table 3 presents the proper sample volume for various BOD ranges. If the sample BOD is expected to be higher than that of the maximum in the table, the sample will require dilution.

3. Insert one of the magnetic stirring bars supplied.

4. Raise a wick about 1/8 inch from the cup floor. Add 4 or 5 drops of potassium hydroxide 45%, Hach Catalog No. 230. CAUTION: Potassium hydroxide is *caustic* and *toxic*.

5. Carefully insert the wick assembly into the brown bottle. *Do not spill any of its contents into the sample*. If any is spilled, start again with a fresh sample.

6. Place the apparatus in the 20° C incubator and set the brown bottle on the apparatus base. Open the screw clamp on the manometer, set the bottle cap loosely on the bottle, and then plug in the stirrer motor.

7. Allow the apparatus to operate about 30 minutes to equalize the temperature, and then close the screw clamp and tighten the cap.

8. Set the manometer scale at zero by loosening the holding screws and sliding the scale until the zero index mark is level with the mercury. If zero cannot be set, open the screw clamp and bottle cap and close again.

9. Repeat steps 1 to 7 for each sample to be tested.

10. Make adequate records—bottle position, sample description, incubator temperature, time started. Gently pinch the short rubber tubing below the clamp momentarily before making a reading. Read the

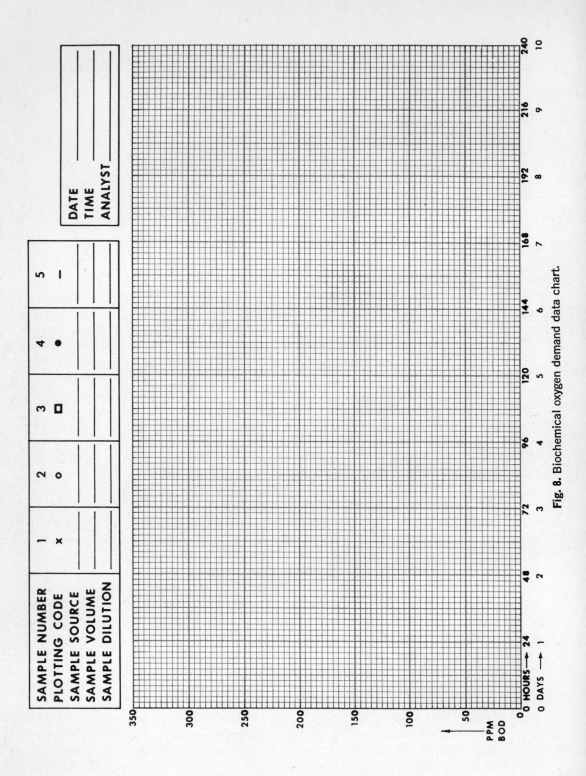

Fig. 8. Biochemical oxygen demand data chart.

80

manometer at least daily and record the readings on the graph (Fig. 8).

11. Remove the sample bottle from the apparatus after the prescribed test period. (Five days for the normal 5-day, 20° C BOD test. However, longer test periods can give additional valuable information.) Record the final results in the result table.

12. If dilutions were made or a sample volume other than 157 ml was taken, multiply the final result by the appropriate factor. (See Table 3. Example: Sample taken, 360 ml; BOD reading after 5 days, 140. Multiplying 140 by 0.2 gives 28 ppm.)

Result table

Samples	5-day BOD	Recommended BOD standard	Remarks
1			
2			
3			
4			
5			

NOTE: Refer to the state and federal standards for the recommended BOD for each of the sample sources.

Questions

Define biochemical oxidation demand (BOD).

What are the state or federal requirements on BOD removal for each of the sample sources?

What materials, organic and inorganic, contribute to the BOD?

Considering the source for each sample collected, list the possible BOD contributors for each sample.

Why is BOD considered an environmental quality parameter from the health standpoint?

Who is responsible for the removal of BOD from wastewater or sewage?

How is BOD removed from wastewater or sewage?

What are some effects of compounds that make up the BOD on plants, animals, and man?

Does a high BOD have an ecological effect on a system? Explain.

PART IV

Physical parameters of water

22
Odor

Materials
1. Collecting bottles with corks or caps
2. Four samples collected from different water sources

Procedure

Each person or group should do a sample.
1. Identify the samples according to number.

1 _____.

2 _____.

3 _____.

4 _____.

2. Store or warm the samples to room temperature or about 78° F.

3. Remove the cap or cork and lightly sniff at the mouth of the bottle. Do not inhale deeply or touch the bottle to your mouth or nose.

4. Determine the odor nature according to Table 4. Post in the result table.

5. Determine the intensity of the odor according to Table 5. Post this in the result table.

6. Collect and post the data in the result table for all samples.

Table 4. Odor characteristics

Code	Nature of odor	Description
A	Aromatic (spicy)	Such as odor of camphor, cloves, lavender, and lemon
Ac	Cucumber	Such as odor of *Synura* (genus of brown, flagellated algae)
B	Balsamic (flowery)	Such as odors of geranium, violets, and vanilla
Bg	Geranium	Such as odor of *Aster onelia*
Bn	Nasturtium	Such as odor of *Aphanizomenon*
Bs	Sweetish	Such as odor of *Coelosphaerium*
Bv	Violets	Such as odor of *Mallomonas*
C	Chemical	Such as odors due to industrial wastes or chemical treatment
Cc	Chlorinous	Odor of free chlorine
Ch	Hydrocarbon	Such as odors of oil refinery wastes
Cm	Medicinal	Such as odors of phenol or iodoform
Cs	Sulfuretted	Odor of hydrogen sulfide
D	Disagreeable	Pronounced unpleasant odors
Df	Fishy	Such as odor of *Uroglenopsis* and *Dinobryon*
Dp	Pigpen	Such as odor of *Anabaena*
Ds	Septic	Such as odor of stale sewage
E	Earthy	Such as odor of damp earth
Ep	Peaty	Such as odor of peat
G	Grassy	Such as odor of crushed grass
M	Musty	Such as odor of decomposing straw
Mm	Moldy	Such as odor of a damp cellar
V	Vegetable	Such as odor of root vegetables

From the Manual of Instruction for Water Treatment Plant Operators, Albany, N. Y., New York State Department of Health.

Table 5

Numerical value	Term	Definition
0	None	No odor perceptible
1	Very faint	An odor that would not be detected ordinarily by the average consumer, but that could be detected in the laboratory by an experienced observer
2	Faint	An odor that the consumer might detect if his attention were called to it, but that would not attract attention otherwise
3	Distinct	An odor that would be detected readily and that might cause the water to be regarded with disfavor
4	Decided	An odor that would force itself upon the attention and that might make the water unpalatable
5	Very strong	An odor of such intensity that the water would be absolutely unfit to drink (a term to be used only in extreme cases)

From the Manual of Instruction for Water Treatment Plant Operators, Albany, N. Y., New York State Department of Health.

Result table

Samples	Nature of odor code	Odor intensity	Safe to drink	Why safe or unsafe
1				
2				
3				
4				

Questions

What samples indicate that the source can be treated so as to be safe to drink? How would you treat them?

What samples indicate the presence of decaying matter? Hypothesize what the source of the decaying matter could be.

Does this odor test indicate the possible presences of harmful chemicals? Bacteria? Give an example of each for each sample.

How does odor reflect the environmental quality of the water source?

23
Turbidity

The American Public Health Association *Standard Methods* states: "Turbidity should be clearly understood to be an expression of the optical property of a sample that causes light to be scattered and absorbed rather than transmitted in straight lines through the sample." Turbidity is caused by the presence of suspended matter, such as clay, mud, algae, silica, rust, bacteria, and calcium carbonate. According to *Standard Methods,* the Jackson Candle Turbidimeter is the standard instrument for the measurement, and the Jackson turbidity unit (JTU) is the standard unit of measurement and expression of turbidity.

Materials

1. Colorimeter, 4445 color filter, 450 nm colorimeter bottles
2. 100 ml graduate cylinder
3. Four samples, each from a lake, stream, pond, and final effluent of a sanitary system

Procedure

Each student or group should do a sample.
1. Identify and number the samples.

   1 _____.

   2 _____.

   3 _____.

   4 _____.

2. Standard preparation—fill a colorimeter bottle with demineralized or distilled water.
3. Insert the turbidity meter scale and 4445 color filter into the meter or use 450 nm.
4. Adjust the light control for a meter reading of zero units.
5. Put 25 ml of the water sample into the colorimeter bottle.
6. Place the colorimeter bottle, with sample, into the meter.
7. Read the Jackson turbidity units and post in the result table.

Result table

Note total solids should be done first, and then post here

Sample	JTU	Total solids
1		
2		
3		
4		

Questions

What does the JTU indicate?

What are the causes of the differences in JTU of each sample?

Why is this test valuable for determining water quality?

Compare the JTU readings with the amount of total solids. Are they significant in relation to one another?

Why are both tests necessary to determine the water quality?

What is the significance of turbidity to the environmental quality of water?

24
Total solids

Material

1. 100 ml porcelain evaporating dishes, 100 ml graduate cylinder sample jars
2. A drying oven, dessicator, gas burner and tripod, analytical balance
3. Four samples, including a sample from a sanitary facility

Procedure

1. Identify the samples according to number.

1 _____.

2 _____.

3 _____.

4 _____.

2. Swirl samples to mix.
3. Weigh the porcelain dish and record in the result table.
4. Pour 100 ml of the sample into the porcelain dish.
5. Put the sample into an oven set at 212° to 218° F or put it on the tripod, using the gas burner to boil gently until sample is evaporated.
6. After evaporation, place the dish, with contents, into the desiccator for cooling and drying.

7. When cool and dry, weigh and record the weights in the result table.
9. Again cool and dry in the desiccator.
10. Weigh and record in the result table.
11. Subtract the weight of the dish determined in step 3 from the weight determined in step 7. This will give the weight of total solids in grams. Post this weight in the result table.
12. Use the following formula to determine ppm of total solids and post in the result table:

$$\text{Weight of total solids in grams} \times \frac{1,000,000}{\text{ml of sample}} = \text{ppm of total solids}$$

13. Determine the volatile total solids. Subtract the weight of that in step 9 from the weight of that in 7 to get the volatile total solids in grams. Determine the ppm with the following formula:

$$\text{Weight of the loss} \times \frac{1,000,000}{\text{ml of sample}}$$
$$= \text{ppm of volatile total solids to step 10}$$

14. Collect the data for the other samples and post in your result table.

Result table

Samples	Dish weight in step 3	Dish weight in step 7	Dish 3 minus dish 7	Weight	ppm of total solids	ppm of volatile total solids	ppm of nonvolatile total solids
1							
2							
3							
4							

Questions

In regard to each sample, what materials compose the total solids?

In regard to each sample, what materials compose the volatile solids? Nonvolatile total solids?

What material of the total solids could be considered as water pollutants? Why?

What material of the volatile solids could be considered as water pollutants? Are there any volatile materials that are beneficial to the water's potability?

How does the quantity and quality of the total solids affect the environmental quality of the water?

PART V

Identification of pesticide residues by thin-layer chromatography

The chlorinated hydrocarbons such as, DDT, aldrin, endrin, heptachlor, and toxaphene are considered persistent pesticides. That is, they do not break down chemically in a short period of time and they are not biodegradable. Therefore, they may persist in the environment for many years. The effects of these pesticides on insects, fowl, animals, and man vary considerably. They may interfere with the conduction of nerve impulses in the central nervous system. Symptoms of their poisoning effect on animals is that of increased excitability, muscular tremors, and convulsions. In addition to the direct physiological effects, most of these pesticides are mutagenic, that is, the causing of mutations within the chromosomes and genes that are responsible for passing on hereditary traits to the progeny. The foregoing effects are those observed in birds and animals that have consumed large quantities of the different pesticides. According to a report by the Secretary of Health, Education, and Welfare, "While there is no evidence to indicate that pesticides presently in use actually cause carcinogenic or teratogenic effects in man, nevertheless, the fact that some pesticides cause these effects in experimental mammals indicates cause for concern and careful evaluation."

The literature dealing with concentrations of pesticides in our ecosystem is confusing and varies considerably. However, almost all animal and human foods are contaminated with pesticides. Practically all terrestrial and aquatic organisms, both invertebrates and vertebrates, have accumulated pesticides within their body tissue.

Many of our edible crops such as potatoes, radishes, carrots, and lettuces have been found to contain pesticide residues at levels from 0.05 to 4 ppm. Lobsters, crabs, shrimp, and fish have been shown to possess high quantities of chlorinated hydrocarbons in their body tissues.

All major rivers and water basins in the United States contain various quantities of DDT, dieldrin, aldrin, and heptachlor. These pesticide residues may be consumed by humans through their drinking water supply. As yet, there are no data to indicate toxic effects to humans from this source. There are, however, data that substantiates the death of millions of aquatic organisms that make up the biotic food chain of the rivers, oceans, streams, etc. The following characteristics and effects of the persistent pesticides are known:

1. They tend to accumulate and concentrate in all plants and animals.
2. They accumulate mainly in fatty tissue.
3. They have been found in all parts of the globe in both the terrestrial and aquatic systems.
4. They are not biodegradable.
5. They have mutagenic effects and interfere with the nervous system.
6. They are presently accumulating in human fat tissue and milk.
7. They have been responsible for the death of thousands of birds and water fowl. Many birds have been found to contain quantities of 100 to 800 ppm of DDT or its metabolites.

8. There are no substantial data to predict possible synergistic or mutagenic effects on humans and their progeny.

9. Millions of aquatic organisms (crustaceans, worms, larvae, and fish) have died from the accumulation of pesticides within their food chain.

10. Many bird and other animal species are facing extinction because of the deleterious effect of the pesticides on their reproductive activities.

The objective of this unit is to utilize thin-layer chromatography for the determination of the presence of pesticide residues in standard solutions, soil, milk, meat, and water.

Thin-layer chromatography

Thin-layer chromatography (TLC) is a separation technique used in chemistry and biological sciences that has been developed as the result of the demand for a simple, rapid method of separating small quantities of related compounds. TLC provides the mechanism to determine the number of pure components in a mixture. Extrapolations can then be made to determine the quantity of each of the components that are present in the material.

The TLC process requires a thin coating of an adsorbent material on a suitable support such as a flexible sheet or glass plate. The mixture to be separated is applied to the adsorbent layer, which is then placed in a vessel containing an appropriate solvent. The solvent transports the applied mixture, which separates into its various components, depending on their relative affinities for the adsorbent as compared to the migrating solvent.

TLC has become widely used as a separation technique, either alone, or in combination with other methods. The simplicity and convenience of operation, the speed, the sharpness of separation, the high sensitivity, and the ease of recovery of the separated components have resulted in its use in medical and clinical studies, production controls, reaction rates, and many molecular identifications.

In TLC, the mechanism of separation may be one of the adsorption, partition, or reversed phase partition, albeit a combination of several mechanisms is usually involved. The results of the migration of the different components are measured as an R_f value computed by dividing the distance the component(s) travel (measured from the origin to the center of the spot it forms on the absorbent layer) by the distance the solvent front travels.

$$R_f = \frac{\text{Distance of spot from origin}}{\text{Distance of solvent front from origin}}$$

It should be noted that R_f values depend on a number of factors, all of which must be controlled if reproducibility of results is required. The following list gives some of the variables that can affect measurements of R_f values:

1. Quality of adsorbent
2. Thickness of adsorbent layer
3. Degree of activity of adsorbent
4. Quality and nature of solvent used
5. Degree of chamber saturation
6. Temperature
7. Running distance
8. Direction of movement (ascending, descending, or horizontal)
9. Amount of sample
10. Impurities present in sample mixtures

In TLC by adsorption, the most commonly operating mechanisms involve the following: The sample is continually fractionated as it migrates through the adsorbent layer. Competition for active adsorbent sites between materials to be separated and the developing solvent produces continuous fractionation. A portion will be adsorbed to the solid adsorbent particles. The various components move different distances. In general, the more polarized compounds are held back by the adsorbent, while the less polarized materials advance farther.

95

25
Identification of pesticides with standard solutions

INSTRUCTOR: The objectives of this experiment are to give the students visual evidence as to the presence of the pesticide residues, and to provide reference chromatograph tables for determining the relative order of migrating distances of the pesticides. These tables can then be used as references when performing the exercises with soil, milk, plant tissue, animal tissue, and water. The following procedure is applicable to the later exercises concerning identification of pesticide residues in soil, milk, plant tissue, animal tissue, and water. Therefore, the extraction methods, introductory statements, distribution graphs, result tables, and questions are given for these exercises. The following companies can supply information and prepared media sheets or plates that can be used for these experiments. However, the experiments are based upon the use of Quanta plates from Quantum Industries.

> Quantum Industries, 341 Kaplin Drive, Fairfield, New Jersey 07006; medium, Quanta plates, type Q1, silica gel medium, 5 × 10 cm or 5 × 20 cm
>
> Gelman Instrument Company, P. O. Box 1448, Ann Arbor, Michigan 48106; medium, Gelman SA sheets, silica gel, 20 × 20 cm or 5 × 20 cm
>
> Eastman Kodak Company, Rochester, New York 14603; medium, Eastman chromagram sheet 6062, alumina for thin-layer chromatography

The indicated companies and many other suppliers can supply TLC pesticide standards, pesticide analysis reagents, and solvents. However, the following sources are convenient for these exercises: Nanogens, Analytical Laboratory and Factory, Box 1025, Watsonville, California 95076. Regis Chemical Company, Chicago, Illinois. Fisher Scientific Company, Chicago, Illinois. Sargent-Welch Scientific Company, Chicago, Illinois. J. T. Baker Chemical Company, Glen Ellyn, Illinois. It is recommended that the references at the end of TLC exercises be checked prior to conducting this exercise.

Procedural preparations

The following should be done prior to the class period:

1. Dilute the standards to 100 ppm in benzene. Use normal dilution procedures.
2. Preparation of the spray reagent: Dissolve 0.1 gram $AgNO_3$ in 1 ml of water. Then add 10 ml of 2-phenoxyethanol. Then add enough acetone to make 200 ml. Then add *one drop* of 30% hydrogen peroxide. Use the indicated sequence when preparing. (Hydrogen peroxide must be stored in a freezer.)
3. Add enough solvent to the developing containers to give a depth of about 1½ cm.
4. Put appropriate quantities of spray reagent into a spraying apparatus.

Materials

For use in identifying standard solutions and pesticides in the extractions made of soil, plant tissue, animal tissue, milk, and water.

1. Aldrin, DDT, dieldrin, heptachlor, heptachlor epoxide or DDD standards diluted to 100 ppm in benzene
2. Solvent, *n*-heptane with 2% acetone (v/v), or hexane
3. 5 or 10 μl micropipettes, applicator
4. Quanta plates, 5 × 20 cm, type Q1 silica gel medium
5. Jars or beakers with minimum dimensions to hold the plates; they are used for developing

6. Ultraviolet light source (short wave)
7. Spraying apparatus
8. Low-pressure air source

Procedure

1. Carefully strip the cover from the Quanta plate.
2. Lay the plate over the outline of the recording table in this exercise.
3. Fill in the pesticides (or extracts) used according to the numbers at the bottom of the outline.
4. While the plate is over the outline, with a pencil, make a small dot on the plate at each of the four dotted locations seen through the plate at the 2 cm height.
5. Make the other indicated notations on the plate.
6. Put the micropipette in the holder and collect a sample of the pesticide solution (or extract) by carefully immersing the pipette 1/8 inch into the solution.
7. Put the sample on the plate at the dot indicated for the sample.
8. This is done by holding the pipette vertical to the plate and lightly touching the surface 4 or 5 times until the sample is adsorbed by the material on the plate. After each application, apply air to the spot until dry.
9. Repeat steps 7 and 8 for each of the pesticides or extracts used.
10. Carefully transfer the plate to the jar containing the developing solvent. Be sure that the lid is returned.
11. Observe the migration of the solvent on the plate until it reaches the 10 cm level that you marked (approximately 15 minutes).
12. When the solvent has reached this level, carefully take the plate out of the jar and let air dry for 5 minutes.
13. When the plate is dry, transfer it to the rack in the vented hood chamber and spray the plate heavily with the spray reagent. Do this 3 times at 1-minute intervals.
14. Let the plate dry in the hood for 10 minutes.
15. After 10 minutes, place the plate under the ultraviolet light source until brown or gray spots are observed (10 to 15 minutes). *Do not* look directly at the ultraviolet light without glasses.
16. After developing, carefully measure the distances the pesticides migrated (as shown by the spots) from the original spot. Post a replicate of the spots, at the measured distance, on the recording table. In some cases, it will be possible to trace the spots by putting the plate under the page containing the recording table.

NOTE: Post your data carefully. This table will be your standard for determining the presence of pesticides in soil, milk, plant tissue, animal tissue, and water. If R_f values are wanted, refer to p. 95.

17. Post in the result table.

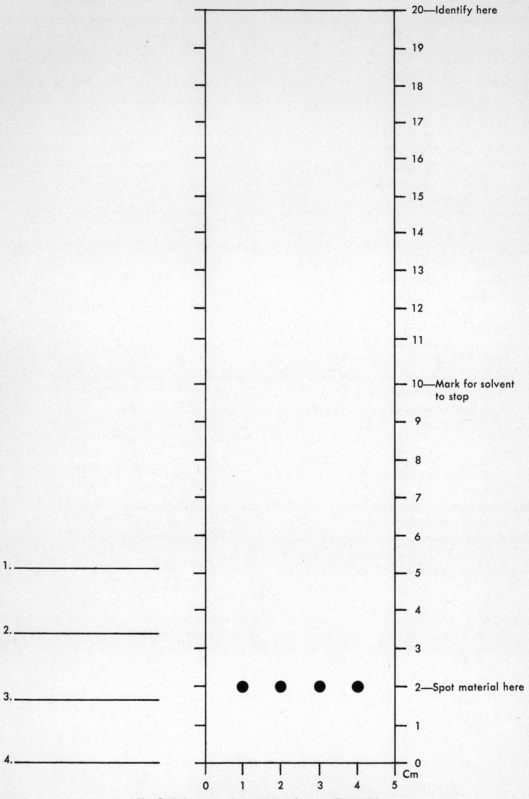

1. _____

2. _____

3. _____

4. _____

20—Identify here

19

18

17

16

15

14

13

12

11

10—Mark for solvent to stop

9

8

7

6

5

4

3

2—Spot material here

1

0

Cm

0   1   2   3   4   5

Fig. 9. Thin-layer chromatograph recording table.

Result table

Pesticide	Distance between original spotting and developed spot	R$_f$ value
Aldrin		
DDT		
Dieldrin		
Heptachlor		
Heptachlor epoxide		

Questions

List the sequence of migration of the pesticides from least movement to greatest movement.

1 _____. 2 _____. 3 _____. 4 _____.

How do these pestcides differ chemically? Check references for this information.

What insects do these pesticides control? Give an insect for each.

What effect do these pesticides have on the insects?

26
Extraction and identification of pesticide residues from soil

INSTRUCTOR: The extraction method for soil is a modification of the Langlois-Stemp-Liska procedure. This procedure is a one-step screening and cleanup method. It will take more than 2 hours to extract and evaporate. Therefore, arrangements should be made for someone to complete the evaporation and the redissolving of the residue. It is suggested that one sample be spiked with a known quantity of pesticide.

PREPARATION OF THE SOIL SAMPLES
Material

1. Methylene chloride (reagent grade), petroleum ether (technical grade), hexane
2. Chromatographic columns, about 20 × 600 mm
3. Glass wool, heating mantle, water bath
4. 50 and 100 ml graduate cylinders, pipettes, 250 and 500 ml beakers, 500 ml Florence flasks, 500 ml graduated separatory funnels
5. Florisil, 60/100 mesh, activated by the supplier at 650° C
6. Concentrators (1000 ml Erlenmeyer flask—standard taper)
7. Ring stand and clamps
8. Anhydrous sodium sulfate (Na_2SO_4)

Procedural preparation

The Florisil must be reactivated at 140° C for 12 to 14 hours. It is then partially deactivated by the addition of 5% water and held in an airtight container for 48 hours before being used.

Prepare the following solvents:

1. 100 ml of 50:50 methylene chloride and petroleum ether solution
2. 1000 ml of 20% methylene chloride petroleum ether solution

Procedure for extracting pesticide residues from soil

1. Mount the chromatograph column vertically on a solid ring stand. Hold in place with 2 clamps.
2. Put a small quantity of glass wool on the inside bottom of the chromatograph column.
3. Put 25 grams of Florisil in the column. Gently tap the column with a spatula while adding the Florisil. Add ½ inch of anhydrous Na_2SO_4.
4. Open the stopcock on the column and wash the Florisil with 50 ml of 50:50 methylene chloride and petroleum ether solvent by elution. Discard the washings.
5. Weigh out 15 grams of the dry soil sample; mix well with 25 grams of Florisil.
6. Add mixture to the column. Gently tap the column while adding the mixture.
7. Mount the 500 ml separatory funnel above the column.
8. Add 650 ml of 20% methylene chloride petroleum ether. Put a concentrator under the column (1000 ml Erlenmeyer flask—standard taper).
9. Open the stopcock and elute with a flow rate of about 5 ml per minute (about 20 drops per 12 seconds).
10. The following pesticides will usually come off in sequence:
    DDT, DDD, DDE, and lindane, 150 ml
    Heptachlor and its epoxide, 250 ml
    Dieldrin, 550 ml
    Endrin, 650 ml
11. Evaporate the eluate to dryness in a water bath at 50° to 60° C or by using the concentrator with the eluant in a heating mantle. This evaporation process must be done under a vented hood.
12. Dissolve the residue in 5 ml of hexane.

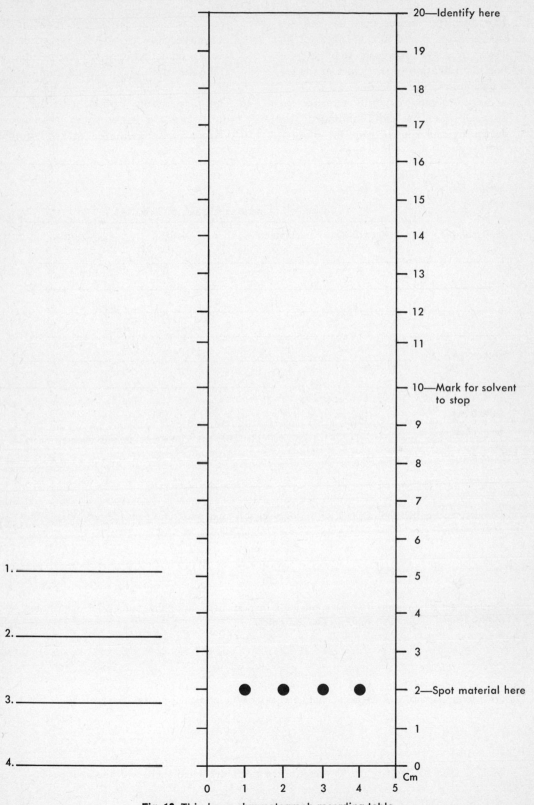

Fig. 10. Thin-layer chromatograph recording table.

13. The dissolved residue may be used immediately or stored in a refrigerator for less than a week, or go to step 14.
14. For identification of pesticides in the extracts, follow the procedure on p. 97.
15. After developing, carefully measure the distance the pesticides migrated (as shown by the spots) from the original spot. Post a replicate of the spots, at the measured distance, on the recording table. In some cases, it will be possible to trace the spots by putting the plate under the page containing the recording table.
16. Compare results (spots) with the standards and post in the result table below. Compare results with the "spiked" sample.

Result table

| Soil sample | Pesticides present or absent, or R_f values | | | |
	DDT or DDD	Heptachlor	Dieldrin	Endrin
1				
2				
3				
4				
Spiked sample				

Questions

Formulate five questions relative to the pesticides that may be found in this extract.

Discuss the significance of pesticide residues in soil as to their effect on the total ecosystem of the area.

Explain the relationship that the application of pesticides has to the presence of chlorinated hydrocarbons in our food supply. Give examples.

Give some physical and chemical characteristics of the pesticide residues that were identified.

Make a table to compare the R_f values of the pesticides found above with those on p. 99.

27
Extraction and identification of pesticide residues from milk

INSTRUCTOR: Milk contains various quantities of compounds, fats, and other organic material that necessitates a *cleanup process* prior to *chromatographic column separation of the pesticide residues*. Therefore, a cleanup process is performed prior to the final chromatographic column extraction. The cleanup process and final extraction have been adapted from the references listed at the end of the exercises. They are based upon Mills, Onley, and Gaither methods. It is recommended that one laboratory period be devoted to cleanup, another laboratory period for final extraction, and then a third laboratory period for the TLC identification, Exercise 24. It is best to spike a sample with a known quantity of pesticide material.

PREPARATION OF THE MILK SAMPLE
Materials—for cleanup

1. 100 ml of whole milk, ice water
2. Blender, 2 ring stands, 3-inch holder
3. 100, 500, and 1000 ml Erlenmeyer flasks, 100 and 500 ml graduate cylinders, 1000 ml separatory funnel, 5 ml volumetric flasks
4. Filter tube with glass wool at the bottom, 25 mm (outer diameter) × 200 mm, funnels, filter paper
5. Concentrators, 500 ml graduate tubes or Florence flasks; water bath apparatus may be used for concentration, bubble column
6. Extraction mixture—1500 ml of acetonitrile, 500 ml of ethyl ether, 500 ml of dioxane, plus 500 ml of acetone; mix thoroughly
7. Petroleum ether (AR grade), 200 ml
8. Anhydrous sodium sulfate (Na_2SO_4)
9. Saturated sodium chloride solution (NaCl)

Cleanup procedure

1. Put 100 ml of whole milk into the blender jar. Add 300 ml of the extraction mixture.
2. Add 100 grams of anhydrous sodium sulfate while mixing at low speed. Then mix at high speed for 2 minutes.
3. Pour this mixture into a 500 ml Erlenmeyer flask. Cool the flask with contents in an ice water bath for 5 to 10 minutes.
4. Filter contents through filter paper into a 500 ml graduate cylinder. Record the volume. (First volume)
5. Put the filtered material (filtrate) into the 1000 ml separatory funnel.
6. Measure 100 ml of petroleum ether into the graduated cylinder used to collect the filtrate. Pour this into the separatory funnel that contains the filtrate.
7. Shake the separatory funnel for 2 or 3 minutes.
8. Then add 10 ml of saturated NaCl solution and 500 ml of distilled water and gently mix.
9. Place the separatory funnel into the ring stand holder. Watch for the material to separate.
10. Drain off the water layer and discard it. The solvent layer will be left.
11. Add 100 ml of distilled water to the separatory funnel and gently shake. Again drain off and discard the water.
12. Repeat step 11.
13. Pour the contents (solvent layer and petroleum ether extract) into the 500 ml graduate cylinder. Record the volume (second volume).
14. Set the 25 × 200 mm filter tube on a stand, add 2 inches of anhydrous sodium sulfate. Put a 500 ml Florence flask or

concentrator (Erlenmeyer flask—standard taper) under the filter tube.

15. Pour the contents of the petroleum ether extract through the column of sodium sulfate.

16. Add 50 ml of petroleum ether to the graduate cylinder used in step 13 and pour this through the column of sodium sulfate.

17. Using the concentrator apparatus, water bath, or nitrogen gas blown gently onto the surface of the extract, concentrate it to about 10 to 20 ml. Do under a vented hood.

18. The concentrate is now ready to be used in the final extraction. Proceed directly or store in a refrigerator till the next laboratory period.

Cleanup volume record

1. First volume from step 4 _____ml
2. Second volume from step 13 _____ml
3. Final volume (estimate) _____ml

FINAL EXTRACTION—THROUGH A CHROMATOGRAPHIC COLUMN
Materials

1. Ethyl ether (AR or USP grade), with 2% (v/v) ethyl alcohol
2. Petroleum ether (AR grade), dioxane (pure grade)
3. Chromatographic column—25 (outer diameter) × 300 mm with Teflon stopcock
4. 60/100 mesh activated Florisil, anhydrous sodium sulfate (Na_2SO_4)
5. Elution mixtures 1 and 2
6. Glass wool, heating mantle, water bath
7. 50 and 100 ml graduate cylinders, 500 ml Florence flasks
8. Concentrators (1000 ml Erlenmeyer flask—standard taper), ring stand, clamps

Procedural information

The Florisil must be reactivated at 140° C for 12 to 14 hours.

Preparation of the elution mixtures

1. Eluting mixture 1: To 1000 ml of petroleum ether add 120 ml of ethyl ether.

2. Eluting mixture 2: To 1000 ml of petroleum ether add 50 ml of ethyl ether and 4 ml of dioxane.

Procedure

1. Clamp the chromatographic column onto a ring stand. Make certain that it is clamped securely and vertical to the rod.

2. Add 4 inches of activated Florisil while gently tapping the column with a spatula.

3. To the top, add about ½ inch of anhydrous sodium sulfate.

4. Place a Florence flask or concentrator (Erlenmeyer flask—standard taper) receiver under the column.

5. Rinse the column with 50 ml of petroleum ether.

6. Pour the concentrated petroleum ether extract (from the cleanup) into the column. Two times, between additions, add 5 ml of petroleum ether.

7. After the extract has gone through the column, add 5 ml more of petroleum ether against the walls of the column.

8. Pour 50 ml of elution mixture 1 into the column at the rate of 5 ml per minute.

9. Pour 175 ml of elution mixture 2 into the column at 5 ml per minute.

10. Using the Florence flask and water bath or the concentrator and heating mantle and using a bubble column, concentrate the collected material to 5 ml under a vented hood.

11. The concentrated extract may be stored til the next laboratory period or you may proceed directly to step 12.

12. For the identification of the pesticide residues in the extract, follow the procedure as given on p. 97.

13. After developing, carefully measure the distance the pesticide residues migrated (as shown by the spots) from the original spotting. Post a replicate of the residue spots at the measured distance, on the thin-layer chromatograph *recording table.* In some cases, it will be possible to trace the spots by putting the plate under the page containing the table.

14. Compare these results (spots) with the

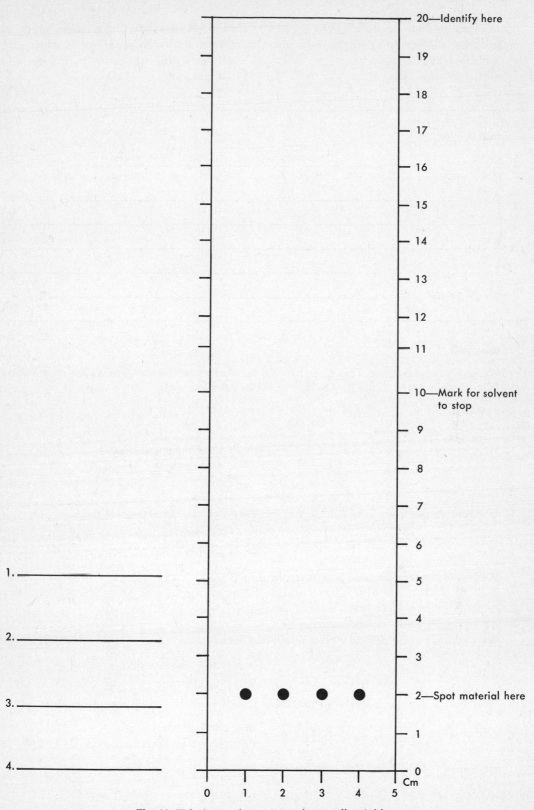

Fig. 11. Thin-layer chromatograph recording table.

standard solution spots of exercise 25. Post the comparison deduction in the result table for this exercise.

15. Collect the data from three other students and post in the result table.
16. Refer to the Appendix, p. 133, for additional data.

Result table

Milk sample	Pesticides present or absent, or R_f values				
	DDT or DDD	Heptachlor	Dieldrin	Endrin	Others
1					
2					
3					
4					
Spiked sample					

Questions

Were there spots on the TLC plate that did not correspond to the spots on the standard solution plates? If so, could they be other pesticide residues?

What pesticide residues were found to be present in each of the milk samples?

Is it possible that there were pesticides in the *milk* that were not extracted? Explain.

If the *milk* is consumed, what happens to the pesticide?

What happens to the pesticides in the forage that the cow eats?

What are the state or federal regulations set up for the presence of these pesticides in milk?

Make a table to compare the R_f values of the pesticides found in this exercise with those on p. 99.

28
Extraction and identification of pesticide residues from plant tissue

INSTRUCTOR: Plant tissue contains various quantities of compounds and organic tissue that necessitates a cleanup process prior to chromatographic column separation of the pesticide residues. Therefore, a *cleanup* process is performed *prior* to the *final chromatographic column extraction*. The cleanup processes have been adapted from the references listed at the end of the exercises. The final extraction procedure is that described by Langlois-Stemp-Lisks. It is recommended that you spike one sample with a known quantity of pesticide.

PREPARATION OF PLANT TISSUE SAMPLES
Materials—for cleanup

1. Plant tissue, about ½ pound
2. Blender, Buchner funnel, 50, 100, and 250 ml graduate cylinders, 1000 ml separatory funnels, 500 ml suction flask
3. 1000 ml acetonitrile (technical grade), 100 ml petroleum ether (AR grade), 50 ml saturated sodium chloride solution, distilled water
4. 10 grams of Celite 545 (Johns-Manville Co., New York)

Cleanup procedures

1. Weigh 100 grams of chopped or blended sample into a blender jar.
2. To this jar add 200 ml of acetonitrile and about 10 grams of Celite.
3. Mix at high speed for 2 minutes.
4. Set up the Buchner funnel apparatus with filter paper and filter the contents into the 500 ml suction flask.
5. Transfer the filtrate to a 250 ml graduated cylinder and *record the volume;* this is the first volume.

6. Transfer the filtrate to a 1000 ml separatory funnel.
7. Put 100 ml of petroleum ether into the same graduate used for the filtrate (in step 5). Pour this into the separatory funnel.
8. Vigorously shake the separatory funnel for 2 minutes.
9. Add 10 ml of saturated NaCl solution and 600 ml of water and mix *gently.*
10. Allow to separate. Drain off the water layer and discard it.
11. Wash the remaining solvent layer two more times with 100 ml of water. Shake gently. Discard the water drained off.
12. Transfer the solvent layer to a 100 ml stoppered graduate. Record the volume (second volume).
13. Add 15 grams of anhydrous Na_2SO_4 and shake vigorously.
14. You are now ready to proceed with the final extraction.

Cleanup volume record

1. First volume from step 5 _____ ml
2. Second volume from step 12 _____ ml
3. Final volume _____ ml

FINAL EXTRACTION
Materials

1. Methylene chloride (reagent grade), petroleum ether (technical grade), hexane
2. Chromatographic columns, about 20 × 600 mm
3. Glass wool, heating mantle, water bath
4. 50 and 100 ml graduate cylinders, pipettes, 250 and 500 ml beakers, 500 ml

Florence flasks, 500 ml graduated separatory funnels, 5 ml volumetric flasks
5. Florisil, 60/100 mesh, activated by the supplier at 650° C
6. Concentrators (Erlenmeyer flask—standard taper)
7. Ring stand and clamps
8. Anhydrous sodium sulfate (Na_2SO_4)

Procedural information

The Florisil must be reactivated at 140° C for 12 to 14 hours. It is then partially deactivated by the addition of 5% water and held in an air tight container for 48 hours before being used.

Prepare the following solvents:

1. 100 ml of 50:50 methylene chloride and petroleum ether solution
2. 1000 ml of 20% methylene chloride petroleum ether solution

Procedure

1. Mount the chromatograph column vertically on a solid ring stand. Hold in place with 2 clamps.
2. Put a small quantity of glass wool on the inside bottom of the chromatograph column.
3. Put 25 grams of Florisil in the column. Gently tap the column with a spatula while adding the Florisil. Add ½ inch of anhydrous Na_2SO_4.
4. Open the stopcock on the column and wash the Florisil with 50 ml of 50:50 methylene chloride and petroleum ether solvent by elution. Discard the washings.
5. Add the *cleanup extract* to the column.

6. Mount the 500 ml separatory funnel above the column.
7. Add 650 ml of 20% methylene chloride petroleum ether. Put a concentrator under the column.
8. Open the stopcock and elute with a flow rate of about 5 ml per minute (about 20 drops per 12 seconds).
9. The following pesticides will usually come off in sequence:
   a. DDT, DDD, DDE, and Lindane, 150 ml
   b. Heptachlor and heptachlor epoxide, 250 ml
   c. Dieldrin, 550 ml
   d. Endrin, 650 ml
10. Evaporate the eluate to dryness in a water bath at 50° to 60° C or by mounting the concentrator with the eluant in a heating mantle. This evaporation process must be done under a vented hood.
11. Dissolve the residue in 5 ml of hexane.
12. The dissolved residue may be used immediately or stored in a refrigerator for less than a week, or go to step 13.
13. For the identification of the pesticides in the extracts, follow the procedure as given on p. 97.
14. After developing, carefully measure the distance the pesticides migrated (as shown by the spots) from the original spot. Post a replicate of the spots, at the measured distance, on the *recording table*. In some cases, it will be possible to trace the spots by putting the plate under the page containing the result table.
15. Compare these results (spots) with the standards and post in the result table for this exercise.

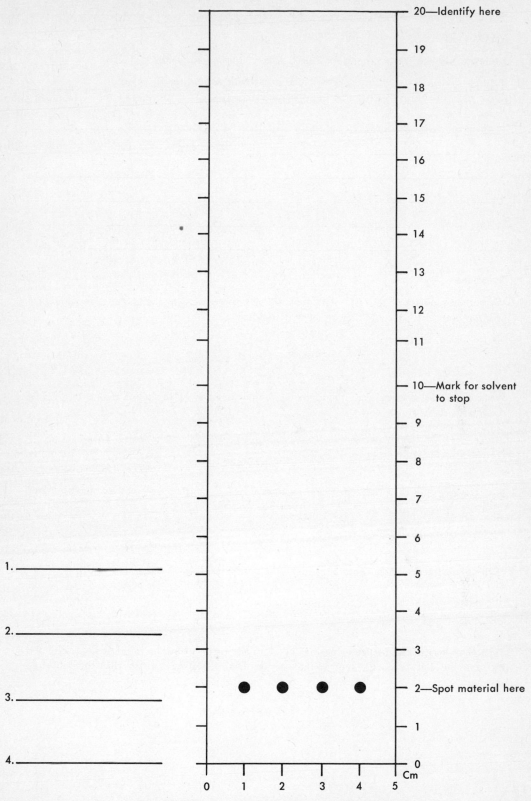

1. _____

2. _____

3. _____

4. _____

20—Identify here

19

18

17

16

15

14

13

12

11

10—Mark for solvent
to stop

9

8

7

6

5

4

3

2—Spot material here

1

0
Cm

0   1   2   3   4   5

Fig. 12. Thin-layer chromatograph recording table.

109

Result table

(Refer to Surface Water Criteria for Public Water Supplies, Appendix, p. 132)

Kind of plant tissue	Pesticides present or absent, or R_f values				
	DDT or DDD	Heptachlor	Dieldrin	Endrin	Others
1					
2					
3					
4					
Spiked sample					

Questions

Were there spots on the TLC plate that did not correspond to the spots on the standard solution plates? If so, could they be other pesticides?

What pesticide residues were found to be present in each of the plant tissues used? List them with the tissue.

Is it possible that there were pesticide residues in the tissue that were not extracted? Why?

If the plant tissue is eaten, what happens to the pesticides?

What happens to the pesticide residues in the plant tissue when the tissue is not consumed by man but left in the field or soil to decay? Describe a food chain that they may go through.

29
Extraction and identification of pesticide residues from animal tissue

INSTRUCTOR: Animal tissue contains various quantities of compounds, fats, and organic material. It is therefore necessary to do a tissue extraction and a cleanup extraction prior to the final chromatographic column extraction. These processses have been adopted from the references listed at the end of the exercises. It is suggested that a laboratory period be devoted to each of the processes—*tissue extraction, cleanup extraction,* and *final extraction*. It is also best to spike a sample with a known pesticide quantity.

PREPARATION FOR PESTICIDE RESIDUES IN ANIMAL TISSUE
Materials for tissue extraction

1. 10 to 20 grams of animal tissue
2. Mortar and pestle
3. Anhydrous sodium sulfate (Na_2SO_4)
4. Centrifuge, refrigerator
5. 50 and 100 ml graduate cylinder, 500 ml beaker or tared flask, 250 ml Erlenmeyer flask
6. Petroleum ether (AR grade)
7. Water bath apparatus, nitrogen gas

Procedure for tissue extraction

1. Identify the animal tissue in the result table. Grind 10 to 20 grams of the animal tissue with anhydrous sodium sulfate. While grinding, add the sodium sulfate (about 100 grams) to absorb the water.
2. Put the mixture into centrifuge bottles.
3. Cool briefly in the refrigerator, then add 100 ml of petroleum ether, and shake vigorously for 1 or 2 minutes.
4. Centrifuge approximately 5 minutes at about 1500 rpm.
5. Pour off the solvent layer into a beaker or tared flask. If the solvent layer is cloudy, filter it through a Buchner funnel with filter paper and 1 inch of sodium sulfate on top of the filter paper.
6. Repeat steps 2 to 5 until all of step 1 mixture has been used.
7. Evaporate the contents at 50° to 60° C, using a water bath under a vented hood to 100 ml.
8. Refrigerate the 100 ml of concentrate or proceed to cleanup.

Materials—for cleanup

1. 100 ml of animal tissue extract, ice water
2. Blender, 2 ring stands, 3-inch holder
3. 100, 500, and 1000 ml Erlenmeyer flasks, 100 and 500 ml graduate cylinders, 1000 ml separatory funnel
4. Filter tube with glass wool at the bottom, 25 mm (outer diameter) × 200 mm, funnels, filter paper
5. Concentrators, 500 ml graduate tubes or Florence flasks; water bath apparatus may be used for concentration, bubble column
6. Extraction mixture—1500 ml of acetonitrile, 500 ml of ethyl ether, 500 ml of dioxane, plus 500 ml acetone; mix thoroughly
7. Petroleum ether (AR grade), 200 ml
8. Anhydrous sodium sulfate (Na_2SO_4)
9. Saturated sodium chloride solution (NaCl)

Cleanup procedure

1. Put 100 ml of animal tissue extract into the blender jar. Add 300 ml of the extraction mixture.
2. Add 100 grams of anhydrous sodium sulfate while mixing at low speed. Then mix at high speed for 2 minutes.

3. Pour this mixture into a 500 ml Erlenmeyer Flask. Cool the flask with contents in an ice water bath for 5 to 10 minutes.
4. Filter contents through filter paper into a 500 ml graduate cylinder. Record the volume (first volume.)
5. Put the filtered material (filtrate) into the 1000 ml separatory funnel.
6. Measure 100 ml of petroleum ether into the graduated cylinder used to collect the filtrate. Pour this into the separatory funnel that contains the filtrate.
7. Shake the separatory funnel for 2 or 3 minutes.
8. Then add 10 ml of saturated NaCl solution and 500 ml distilled water, and gently mix.
9. Place the separatory funnel into the ring stand holder. Watch for the material to separate.
10. Drain off the water layer and discard it. The solvent layer will be left.
11. Add 100 ml of distilled water to the separatory funnel and gently shake. Again drain off and discard the water.
12. Repeat step 11 again.
13. Pour the contents (solvent layer and petroleum ether extract) into the 500 ml graduate cylinder. Record the volume (second volume).
14. Set up the 25 × 200 mm filter tube on a stand and add 2 inches of anhydrous sodium sulfate. Put a 500 ml Florence flask or concentrator under the filter tube.
15. Pour the contents of the petroleum ether extract through the column of sodium sulfate.
16. Add 50 ml of petroleum ether to the graduate cylinder in step 13 and pour this through the column of sodium sulfate.
17. Using the concentrator, water bath, or nitrogen gas blown gently onto the surface of the extract, concentrate it to about 10 to 20 ml. Do under a vented hood.
18. The concentrate is now ready to be used in the final extraction. Proceed directly or store in a refrigerator till the next laboratory period.

Cleanup volume record

1. First volume from step _____ml
2. Second volume from step 13 _____ml
3. Final volume (estimate) _____ml

Materials for final extraction—through a chromatographic column

1. Ethyl ether (AR or USP grade), with 2% (v/v) ethyl alcohol
2. Petroleum ether (AR grade), dioxane (pure grade)
3. Chromatographic column, 25 (outer diameter) × 300 mm with Teflon stopcock
4. Activated Florisil, 60/100 mesh; anhydrous sodium sulfate (Na_2SO_4)
5. Elution mixtures 1 and 2
6. Glass wool, heating mantle, water bath
7. 50 and 100 ml graduate cylinders, 500 ml Florence flasks
8. Concentrators, ring stand, clamps

Procedural information

The Florisil must be reactivated at 1400° C for 12 to 14 hours. It is then partially deactivated by the addition of 5% water and held in an airtight container for 48 hours before being used.

Preparation of the elution mixtures

1. Eluting mixture 1: To 1000 ml of petroleum ether add 120 ml of ethyl ether.
2. Eluting mixture 2: To 1000 ml of petroleum ether add 50 ml of ethyl ether and 4 ml of dioxane.

Procedure

1. Clamp the chromatographic column onto a ring stand. Make certain that it is clamped securely and vertical to the rod.
2. Add 4 inches of activated Florisil while gently tapping the column with a spatula.
3. To the top, add about ½ inch of anhydrous sodium sulfate.
4. Place a Florence flask or concentrator receiver under the column.
5. Rinse the column with 50 ml of petroleum ether.
6. Pour the concentrated petroleum ether

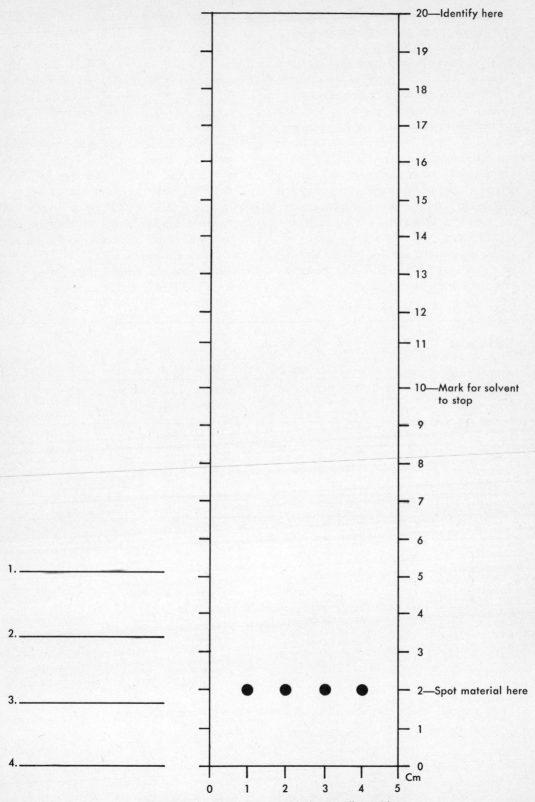

1._____

2._____

3._____

4._____

20—Identify here

19

18

17

16

15

14

13

12

11

10—Mark for solvent
to stop

9

8

7

6

5

4

3

2—Spot material here

1

0

Cm

0 1 2 3 4 5

Fig. 13. Thin-layer chromatograph recording table.

extract (from the cleanup) into the column. Two times between additions, add 5 ml of petroleum ether.

7. After the extract has gone through the column, add 5 ml more of petroleum ether against the walls of the column.

8. Pour 50 ml of elution mixture 1 into the column at the rate of 5 ml per minute.

9. Pour 175 ml of elution mixture 2 into the column at 5 ml per minute.

10. Using the Florence flask and water bath or the concentrator and heating mantle and using a bubble column, concentrate the collected material to 5 ml under a vented hood.

11. The concentrated extract may be stored till the next laboratory period or you may proceed directly to step 12.

12. For the identification of the pesticide residues in the extract, follow the procedure as given on p. 97.

13. After developing, carefully measure the distance the pesticide residues migrated (as shown by the spots) from the original spotting. Post a replicate of the residue spots, at the measured distance, on the thin-layer chromatograph *recording table*. In some cases, it will be possible to trace the spots by putting the plate under the page containing the table.

14. Compare these results (spots) with the standard solution spots of Exercise 24. Post the comparison deduction in the *result table* for this exercise.

15. Collect the data from three other students and post in the result table.

Result table

Animal tissue sample	Pesticides present or absent, or R_f values				
	DDT or DDD	Heptachlor	Dieldrin	Endrin	Others
1					
2					
3					
4					
Spiked sample					

Questions

Were there spots on the TLC plate that did not correspond to the spots on the standard solution plates? If so, could they be other pesticide residues?

What pesticide residues were found to be present in each of the animal tissue samples?

Is it possible that there were pesticides in the animal tissue that were not extracted? Explain.

If the animal tissue is consumed, what happens to the pesticide?

What happens to the pesticides in the forage that the animal eats?

What are the state or federal regulations set up for the presence of these pesticides in animal tissue?

30
Extraction and identification of pesticide residues from water

INSTRUCTOR: The concentrations of pesticide residues vary with the time of the year. It also varies with the geographical location of the water source. Therefore, to ensure the success of the laboratory exercise, it is best to spike one water sample with a known quantity of mixed pesticides. The thin-layer chromatographic results can be used as a standard for the comparison of the other water extracts.

Materials

1. Water sample, 800 to 900 ml
2. 1000 ml separatory funnel
3. Concentrating apparatus (1000 ml Erlenmeyer flask—standard taper), fluidized sand bath, bubble column
4. 125 and 150 Erlenmeyer flasks, 5 and 10 ml volumetric flasks, 25 and 50 ml graduate cylinders, disposable 1 ml pipettes
5. Anhydrous sodium sulfate (Na_2SO_4)
6. 100 ml n-hexane
7. Ring stand, clamps, ring clamp

Procedural information

Water samples are extracted with n-hexane in such a way that the water, or the water sediment mixture, and the container itself are exposed to the solvent by the following technique; it is necessary to make at least three extractions:

Procedure

1. Pour the water sample into a 1-liter separatory funnel.
2. Add 25 ml of hexane to the empty sample bottle and swirl.
3. Pour the hexane into the separatory funnel.

4. The hexane that remains on the sides of the sample bottle should be rinsed into the separatory funnel with small portions of the water sample.
5. Shake the funnel for 1 minute to mix thoroughly.
6. Put the separatory funnel on the ring stand for 10 minutes to allow the contents to separate.
7. Drain off the aqueous layer into the original sample bottle.
8. If the hexane layer becomes emulsified, add small amounts of distilled water and shake intermittently to break the emulsion.
9. Pour the hexane layer from the top of the separatory funnel into a 125 ml Erlenmeyer flask that contains about 0.5 gram of anhydrous sodium sulfate.
10. Collect the contents of step 9 from three other samples into a 125 ml flask containing 0.5 gram of sodium sulfate.
11. Decant the combined extracts quantitatively into a concentrating apparatus.
12. Remove the hexane by heating on a fluidized sand bath at 100° C until about 25 ml of the extract remains.
13. Add 1 gram of anhydrous sodium sulfate and mix thoroughly.
14. The concentrated extract may be stored till the next laboratory period or you may go directly to step 15.
15. For the identification of the pesticide residues in the extract, follow the procedures as given on p. 97.
16. After developing the TLC plate, carefully measure the distance the pesticide residues migrated (as shown by the spots) from the original spotting. Post a replicate of the residue spots, at the measured

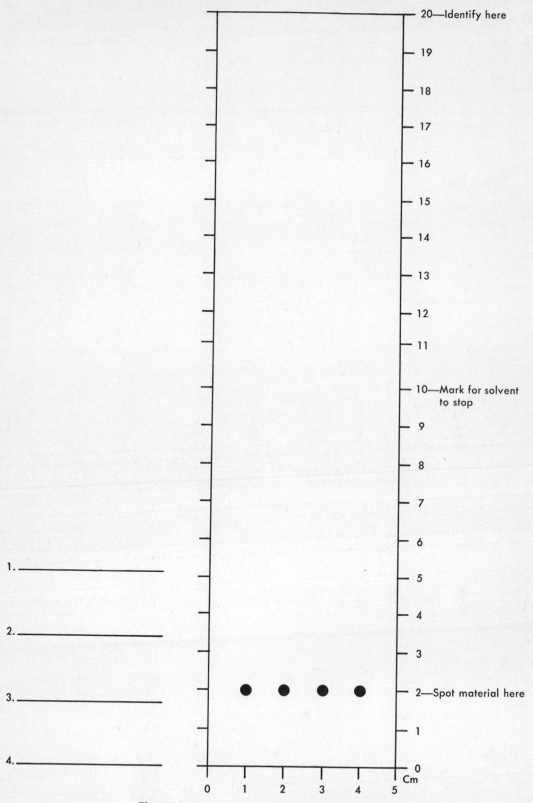

Fig. 14. Thin-layer chromatograph recording table.

117

distance, on the thin-layer chromatograph *recording table*. In some cases, it will be possible to trace the spots by putting the plate under the page containing the table.

17. Compare these results (spots) with the standard solution (Exercise 25) or the spiked sample spots. Post the pesticides identified in the *result table* of this exercise.

18. Collect the data from three other samples tested and post in the result table.

19. Refer to the Introduction and Appendix, for additional information on pesticides.

Result table

Sample	Pesticides present or absent, or R_f values				
	DDT or DDD	Heptachlor	Dieldrin	Endrin	Others
1					
2					
3					
4					
Spiked sample					

Questions

Were there spots on the TLC plate that did not correspond to the spots on the standard solution or spiked sample plates? If so, could they be other pesticide residues?

Is it possible that there were other pesticides in the water that were not extracted? Explain.

If the water is consumed, what happens to the pesticides?

Describe the accumulation of the pesticides in an aquatic food chain. How does this affect man?

How can we control the addition of pesticides in our lakes, streams, and rivers?

What are the state and federal regulations set up for the presence of each of the pesticide residues in water sources?

References for PART V

Abbott, D. C., and J. Thomson. 1965. The application of thin-layer chromatographic techniques to the analysis of pesticide residues. Residue Reviews, vol. II.

Langlois, B. E., A. R. Stemp, and B. J. Liska. 1964. Insecticide residues, rapid clean-up of dairy products for analysis of chlorinated insecticides by E. C. and G. C., J. Agr. Food Chem. 17(3):243-245.

Mills, P. A., J. A. Onley, and R. A. Garther. 1963. Rapid method for chlorinated pesticide residues in nonfatty foods. J.A.O.A.C. 46(2): 186-191.

Onley, J. H. 1964. Rapid method for chlorinated pesticide residues in fluid milk. J.A.O.A.C. 47 (2):317-321.

United States Department of Health, Education, and Welfare. Revised January 1969. Food and drug, pesticide analytical manual. Vol. I, sections 200, 210, 220, 400, and 610. U. S. Government Printing Office, Washington, D. C.

PART VI

Appendixes

A
Glossary

Aerobacter aerogenes A genus and species of bacteria included in the coliform group that may indicate pollution of old origin or the result of growth such as that sometimes occurring in leather washers, jute packing, and wood.

aerobic bacteria Bacteria that require free oxygen for their metabolic processes.

aerobic digestion Digestion of organic matter by means of aeration.

aerosol A suspension of fine solid or liquid particles in air or gas, as smoke, fog, or mist.

agglomeration Coalescence of dispersed suspended matter into larger flocs or particles causing them to settle.

air Mixture of gases that surrounds the earth and forms its atmosphere, composed primarily of oxygen and nitrogen plus a heterogenous group of substances such as carbon dioxide, carbon monoxide, some water vapor, other gases, and suspended particles.

algae One-celled or many-celled plants, usually aquatic, and capable of carrying on photosynthesis.

algal bloom Large masses of microscopic and macroscopic plant life, mostly green algae, occurring in bodies of water.

algicide Any substance or chemical that will kill or control algal growths.

alkali Any of the soluble salts, such as sodium, potassium, magnesium, and calcium, that have the property of combining with acids to form neutral salts.

alkaline Condition of water or soil that contains a sufficient amount of alkali substances to give a pH above 7.0.

alkaline water Water with a pH greater than 7.0 because of the large amount of OH$^-$ ions.

alkalinity Condition imparted by the water's content of carbonates, bicarbonates, hydroxides, and occasionally borates, silicates, and phosphates. Usually expressed in milligrams per liter of equivalent calcium carbonate.

ameba A macroscopic one-celled animal, a protozoan.

amino acid Any of a class of organic compounds that contain one or more carboxyl groups and one or more amino groups. The $RCH(NH_2)$—COOH are the building blocks from which proteins are synthesized.

anaerobic bacteria Bacteria that can carry on their metabolic processes in the absence of oxygen.

bacteria Unicellular microscopic organisms lacking chlorophyll. Bacteria are usually spheroid, rod-like, or curved in shape, but many grow in colonies as chains or branching filaments or in clumps.

biochemical (As adjective) A chemical change resulting from biological action. Pertaining to the chemistry of plant and animal metabolism. (As noun) A chemical compound produced by fermentation.

biochemical oxygen demand (BOD) A measure of the quantity of oxygen required for the biochemical oxidation of organic matter, in water, during a specified time, at a specified temperature, and under specified conditions. A standard test used in assessing polluted water. The amount of oxygen used by microorganisms during oxidation.

biodegradation (biodegradability) Breaking down or mineralization of natural or synthetic organic materials by the microorganisms in soils, natural bodies of water, or wastewater treatment systems.

biological oxidation A metabolic process whereby living organisms utilize oxygen to convert the organic matter contained in wastewater into a more stable or a mineral form.

biota Animal and plant life of a stream or other water body.

centigrade A thermometer temperature scale in which 0 degrees marks the freezing point and 100 degrees the boiling point of water, at 760 mm of mercury barometric pressure. Also called the Celsius scale. To convert temperature on this scale to Fahrenheit, multiply by 9/5 and add 32.

chemical oxygen demand (COD) A measure of the oxygen-consuming capacity of inorganic and organic matter found in water or polluted water. Expressed as the amount of oxygen consumed from a chemical oxidant. Does not differentiate between stable and unstable organic matter and does not show a correlation with biochemical oxygen demand. Also known as OC and DOC, oxygen consumed and dichromate oxygen consumed, respectively.

chloramines Compound of inorganic or organic nitrogen and chlorine.

123

chlorination Application of chlorine to water or wastewater to kill bacteria and fungi, and also to accomplish other biological or chemical results.

chlorine An element ordinarily existing as a greenish yellow gas about 2.5 times as heavy as air. At atmospheric pressure and a temperature of −30.1° F, the gas becomes an amber liquid that is about 1.5 times as heavy as water. The chemical symbol of chlorine is Cl, its atomic weight is 25.457, and its molecular weight is 70.914.

chlorinated hydrocarbons Compounds that contain chlorine, hydrogen, and carbon, such as dieldrin, DDT, and endrin. They affect the central nervous system of animals.

combustion Act or process of burning.

Crenothrix A genus of bacteria characterized by unbranched attached filaments having a gelatinous sheath in which iron is deposited. These organisms precipitate iron, which is deposited in pipes and thus reduces carrying capacity. They produce color in water and after their death impart a disagreeable taste to water.

cubic meter Common measurement of air volume.

degradation Separating of compounds and substances by biological action.

degree On the centigrade thermometer scale, $\frac{1}{100}$ of the interval from the freezing point to the boiling point of water under standard conditions; on the Fahrenheit scale, $\frac{1}{180}$ of this interval.

dehydration Chemical or physical process whereby water that is combined chemically or physically with other matter is removed.

deoxygenation Loss of the dissolved oxygen in a liquid either under natural conditions through biochemical oxidation or organic matter or by chemical reducing agents.

detergent A group of synthetic, organic, liquid, or water-soluble cleaning agents that are inactivated by hard water and have wetting-agent and emulsifying-agent properties. Unlike soap, they are not manufactured from fats and oils, but they clean like soap.

dissolved solid Residues of the dissolved constituents in water. May be composed of many different anhydrous residues.

ecology Branch of biology dealing with the interrelationships between organisms and their environment, including man. How the life processes of one group affect the population of another.

effluent Liquid substance (water) that flows out of a containing space, such as a treatment facility.

Enterobacter, see **Aerobacter**

enterococci A group of cocci normally inhabiting the intestines of man or animals.

enzyme An organic catalyst produced by living cells. All enzymes are proteins, but not all proteins are enzymes.

Escherichia coli (E. coli) A species of *Escherichia* bacteria that inhabit the intestine of man and all vertebrates and are included in the coliform group. Their presence shows evidence of fecal contamination.

exhaust The used fuel that escapes from an engine.

Fahrenheit The temperature scale in which the level 32° marks the freezing point and 212° the boiling point of water at 760 mm of mercury of barometric pressure. To convert to centigrade, subtract 32 and multiply by 5/9.

fauna The animals of a particular habitat.

fermentation A change brought about by ferment, as yeast enzymes. Changes in organic matter or organic wastes brought about by microorganisms.

flora The plant population of a particular habitat.

fog Chemical combination that cuts down vision.

fungi Small non–chlorophyll bearing plants that lack roots, stems, or leaves, occur (among other places) in water, wastewater, or wastewater effluents, and grow best in the absence of light. Their decomposition after death may cause disagreeable tastes and odors in water; in some wastewater treatment processes they are helpful, and in others they are detrimental.

hardness Condition of water, caused by salts of calcium, magnesium, and iron such as bicarbonates, carbonates sulfates, chlorides, and nitrates.

heterotrophic organisms Bacteria that propagate and metabolize organic materials.

household detergent Detergents produced for use in home laundries. They include anionic or nonionic surfactants plus phosphates and many minor ingredients such as perfumes, brighteners, and bleaches.

hydrocarbon Any compound consisting solely of carbon and hydrogen.

inorganic matter Chemical substances of mineral origin, excluding carbon as the basic structure.

inversion A condition of a downward stream of particles and gases caused by the formation of an atmospheric ceiling of cold and warm air, with resultant increase in concentration of suspended particulate and gases.

irritant That which makes one unnaturally sensitive or sore.

microgram and milligram Weight measurements of material found in air, both solid and gaseous.

most probable number (MPN) That number of organisms per unit of volume that, in accordance with statistical theory, would be more likely than any other number to yield the observed test result or that would yield the observed test result with the greatest frequency. Expressed as density of organisms per 100 ml. The results are computed from the number of positive findings of the coliform group of organisms resulting from multiple-portion decimal-dilution cultures.

organic matter Animal- or vegetable-produced chemical substances of basically carbon structure, comprising compounds consisting of hydrocarbons and their derivatives.

organic phosphorus compounds Compounds that are derivatives of phosphoric acid. They are toxic to animals in that they interfere with cholinesterase enzyme activity.

oxidation Addition of oxygen to a compound. Any reaction that involves the loss of electrons from an atom.

particle size Relative size of the average particles in dust samples. Measured in units known as microns.

particulate matter Solid material released as a result of a combustion process or natural dust.

parts per million (ppm) The number of weight or volume units of constituent present with each 1 million units of the major constituent of a solution or mixture. 1 mg per liter.

pesticide Any substance or chemical applied to kill or control weeds, insects, algae, rodents, and other undesirable pests (chlorinated hydrocarbons and organic phosphate compounds are examples of pesticides).

photochemical Of or having to do with the chemical action of light.

photosynthetic bacteria Bacteria that obtain their energy for metabolism from light photosynthesis.

plate count Number of colonies of bacteria grown on certain solid media at a given temperature and incubation period, normally expressed in number of bacteria per milliliter of sample.

pollution Condition caused by the presence of harmful or objectionable material in water or air.

primary treatment The first major treatment in a wastewater treatment works, usually sedimentation. The removal of a suspended matter but little or no colloidal and dissolved matter.

saprophytic bacteria Bacteria that live on dead organic matter.

smog Combination of smoke and fog.

smoke Gas and particulate matter.

standard biochemical oxygen demand Biochemical oxygen demand as determined under standard laboratory procedure for 5 days at 20° C, usually expressed in milligrams per liter.

Staphylococcus A genus of spherical bacteria, occurring in pairs, tetrads, and irregular clusters. Some species are pathogenic to man.

steam Gas phase of a liquid.

total solids The quantity of dissolved and undissolved constituents in water or wastewater, usually stated in milligrams per liter.

turbidimeter An instrument for measurement of turbidity, in which a standard suspension is used for reference.

turbidity A condition in water or wastewater caused by the presence of suspended material resulting in the scattering and absorption of light rays. An analytical quantity reported in arbitrary turbidity units determined by measurements of light diffraction.

ultraviolet Invisible rays of light, just beyond the violet in the visible spectrum.

urban Of or having to do with cities or towns.

water A transparent, odorless, tasteless liquid, that is a compound of hydrogen and oxygen (H_2O), freezing at 32° F or 0° C and boiling at 212° F or 100° C and constitutes rain, oceans, lakes, rivers, and other such bodies; it contains 11.188% hydrogen and 88.812% oxygen, by weight. It may exist as a solid, liquid, or gas, is normally found in the lithosphere, hydrosphere, and atmosphere, and may have other solid, gaseous, or liquid materials in solutions or suspension.

B

Weight and volume equivalents

1000 ppm	100 ppm	1 ppm
1 gram per 1000 ml	100 mg per 1000 ml	1 mg per 1000 ml
1000 mg per 1000 ml	0.1 mg per 1 ml	1 μg per 1 ml
1 mg per 1 ml	0.1 μg per 1 μl	0.001 μg per 1 μl
1 μg per 1 μl	100 ng per 1 μl	1 ng per 1 μl
1000 ng per 1 μl		

Weight units	Volume units	Length units
1 kg = 1000 grams	1 ft^3 = 28,317 cm^3	1 m = 39.37 inches
1 gram = 0.001 kg	1 m^3 = 10^6 cm^3	1 m = 0.001 cm
1 gram = 0.002 lb		1 cm = 10 mm
1 gram = 1000 mg		1 mm = 1000 microns
1 mg = 0.001 gram		1 micron = 0.000039 inch
1 mg = 1000 gram		

C

Effects of particulate matter

Concentrations	Effect
mg/m³	
25 to 50	Background levels
75 to 100	Considered satisfactory air quality by most people
over 100	Increased mortality from all causes, increased mortality from chronic respiratory diseases
150 to 200	Considered dirty by most people
over 200	Considered excessively dirty by most people
100	Visibility reduced to about 5 miles
140	Visiblllty reduced to about 3 miles at a humidity of 60%
Cohs/1000 linear ft	
over 0.3	Increase in total morbidity and incidence of cardiovascular diseases among middle-class individuals 55 years of age or older
	Increased mortality from respiratory diseases among the middle socioeconomic class
0.4 at 60% relative humidity	Visibility reduced to about 9 miles
0.4 at 90% relative humidity	Visibility reduced to about 3 miles
2.6 at 60% relative humidity	Visibility reduced to about 3 miles
tons/mile²/month	
5	Background level
15	Considered satisfactory for residential areas
10 3-month average above background areas except those zoned heavy industrial	
25 3-month average above background in zoned heavy industrial areas	
30	Considered dirty by most people

McGuiness, B. J. 1968. Problems of air pollution. Manuscript presented at Purdue University, Indianapolis, Indiana.

D

Effects of sulfur dioxide

Sulfur dioxide concentration, ppm	Exposure period and effect	Measurement methods
	Yearly exposure	
Trace	Metal corrosion begins	PbO_2 candle
0.01 to 0.02	Significant metal corrosion	PbO_2 candle
	Impaired pulmonary function	PbO_2 candle, West-Gaeke
	Increased cardiovascular morbidity	PbO_2 candle, West-Gaeke
0.02 to 0.03	Increased respiratory death rates for area studied	PbO_2 candle
	Detectable chronic injury to perennial vegetation	Thomas autometer
	2- to 4-day exposure	
0.07 to 0.25	Hospital admissions for cardiorespiratory diseases increase	H_2O_2
0.20 to 0.30 for 3 days	Rhinitis, sore throat, cough, and eye-irritation rates increased	By electroconductivity
0.20 to 0.86 for 3 days	Cardiorespiratory mortality increased	H_2O_2
	Acute vegetation injury	Pure gas
	24-hour exposure	
0.21	Bronchitic patients' health deteriorates	H_2O_2
0.25	Increased total death rates	H_2O_2
0.28	Detectable injury to sensitive vegetation	Pure gas
	Brief exposures	
0.04	Visibility reduced to 10 miles at 70% relative humidity	Calculated effect
0.08	Cortical conditioned reflexes produced; repeated 10-second exposures	(No method indicated)
0.10	Visibility reduced to 4 miles at 70% humidity	Calculated effect
0.30	Taste threshold	Pure gas
0.50	Visibility reduced to 0.85 mile at 70% humidity	Calculated effect
0.5 for 1 sec	Odor threshold	Pure gas
0.5 for 4 hr	Detectable injury to sensitive vegetation	Pure gas
0.5 for 7 hr	Acute injury to trees and shrubs	Pure gas
1.0 for 10 min	Respiration and pulse rates increase	Pure gas
1.6 for 1 to 5 min	Threshold for inducing measurable bronchoconstriction in healthy people	Pure gas

McGuiness, B. J. 1968. Problems of air pollution. Manuscript presented at Purdue University, Indianapolis, Indiana.

E

Effects of nitrogen dioxide

Concentration, ppm	Exposure time	Effect
0.1		Limit of acceptability for coloration effect in aerosol-free air with viewing distance of 10 miles
0.25		Limit of acceptability for coloration effect in normal metropolitan area air with viewing distance of 10 miles when the visibility is 20 miles
0.5	3 months	Increased susceptibility to infection in mice by certain aerosolized bacteria
<0.5	12 to 19 days	Significant growth reduction in tomato and bean seedlings; no visible lesion damage; chlorosis of leaves reported
1 to 3		Odor threshold
2.5	7 hours or more	Bean, tomato, and *Ricotiana glutinosa* leaves damaged, with white lesions occurring
3	4 to 8 hours	Pinto bean leaf damage
3.5	2 hours	Increased susceptibility to infection in mice by certain aerosolized bacteria
13		Nasal and eye irritation noticeable

McGuiness, B. J. 1968. Problems of air pollution. Manuscript presented at Purdue University, Indianapolis, Indiana.

F

Effects of carbon monoxide*

Concentration, ppm	Exposure time	Effects
5	20 minutes	Reflex changes in the higher nerve centers
5 to 10		Average levels of CO in St. Louis and most large cities
30	8 hours or more	Impairment of visual and mental acuity (5% carboxyhemoglobin)
70 to 100		Maximum levels occurring in some large cities
200	2 to 4 hours	Tightness across the forehead, possible slight headache
500	2 to 4 hours	Severe headache, weakness, nausea, dimness of vision, possibility of collapse
1000	2 to 3 hours	Rapid pulse rate, coma with intermittent convulsions, and Cheyne-Stokes respiration
2000	1 to 2 hours	Death

*All the effect levels pertain to healthy individuals. Specific effect levels for individuals who for other reasons are approaching the levels of tolerability are not available.

McGuiness, B. J. 1968. Problems of air pollution. Manuscript presented at Purdue University, Indianapolis, Indiana.

G

Air quality standards (by the Environmental Protection Agency)

Sulfur oxides
Primary
    80 micrograms per cubic meter (0.03 ppm)—annual arithmetic mean.
    365 micrograms per cubic meter (0.14 ppm)—maximum 24-hour concentration not to be exceeded more than once per year.
Secondary
    60 micrograms per cubic meter 0.02 ppm)—annual arithmetic mean.
    260 micrograms per cubic meter (0.1 ppm)—maximum 24-hour concentration not to be exceeded more than once per year, as a guide to be used in assessing implementation plans to achieve the annual standard.
    1300 micrograms per cubic meter (0.5 ppm)—maximum 3-hour concentration not to be exceeded more than once per year.

Particulate matter
Primary
    75 micrograms per cubic meter—annual geometric mean.
    260 micrograms per cubic meter—maximum 24-hour concentration not to be exceeded more than once per year.
Secondary
    60 micrograms per cubic meter—annual geometric mean, as a guide to be used in assessing implementation plans to achieve the 24-hour standard.
    150 micrograms per cubic meter—maximum 24-hour concentration not to be exceeded more than once per year.

Carbon monoxide (primary and secondary)
    10 milligrams per cubic meter (9 ppm)—maximum 8-hour concentration not to be exceeded more than once per year.
    40 milligrams per cubic meter (35 ppm)—maximum one hour concentration not to be exceeded more than once per year.

Photochemical oxidants (primary and secondary)
    160 micrograms per cubic meter (0.08 ppm)—maximum one hour concentration not to be exceeded more than once per year.

Hydrocarbons (primary and secondary)
    160 micrograms per cubic meter (0.24 ppm)—maximum 3-hour concentration 6:00 to 9:00 a.m.) not to be exceeded more than once per year.

Nitrogen dioxide (primary and secondary)
    100 micrograms per cubic meter (0.05 ppm)—annual arthmetic mean.

H

Surface water criteria for public water supplies

Two types of criteria are defined as follows:

1. Permissible criteria—Those characteristics and concentrations of substances in raw surface waters that will allow the production of a safe, clear, potable, esthetically pleasing, and acceptable public water supply that meets the limits of drinking water standards after treatment. This treatment may include, but will not include more than, the processes described above.
2. Desirable criteria—Those characteristics and concentrations of substances in the raw surface waters that represent high-quality water in all respects for use as public water supplies. Water meeting these criteria can be treated in the defined plants with greater factors of safety or at less cost than is possible with waters meeting permissible criteria.

Several words used in the table require explanation in order to convey the subcommittee's intent:

Narrative—The presence of this word in the table indicates that the subcommittee could not arrive at a single numerical value that would be applicable throughout the country for all conditions.

Absent—The most sensitive analytical procedure in *Standard Methods for the Examination of Water and Wastewater* (or other approved procedure) does not show the presence of the subject constituent.

Virtually absent—This term implies that the substance is present in very low concentrations and is used where the substance is not objectionable in these barely detectable concentrations.

SURFACE WATER CRITERIA FOR PUBLIC WATER SUPPLIES

Constituent or characteristic	Permissible criteria	Desirable criteria
Physical		
Color (color units)	75	<10
Odor	Narrative	Virtually absent
Temperature*	Narrative	Narrative
Turbidity	Narrative	Virtually absent
Microbiological		
Coliform organisms	10,000/100 ml†	<100/100 ml†
Fecal coliforms	2000/100 ml†	< 20/100 ml†
	(mg/L)	(mg/L)

*The defined treatment process has little effect on this constituent.

†Microbiological limits are monthly arithmetic averages based on an adequate number of samples. Total coliform limit may be relaxed if fecal coliform concentration does not exceed the specified limit.

‡As parathion in cholinesterase inhibition. It may be necessary to resort to even lower concentrations for some compounds or mixtures.

From *Report of the Committee on Water Quality Criteria*, April 1, 1968, Federal Water Pollution Control Administration, Washington D. C. p. 20.

SURFACE WATER CRITERIA FOR PUBLIC WATER SUPPLIES—cont'd

Constituent or characteristic	Permissible criteria	Desirable criteria
Inorganic chemicals		
Alkalinity	Narrative	Narrative
Ammonia	0.5 (as N)	<0.01
Arsenic*	0.05	Absent
Barium*	1	Absent
Boron*	1	Absent
Cadmium*	0.01	Absent
Chloride*	250	<25
Chromium, hexavalent*	0.05	Absent
Copper*	1	Virtually absent
Dissolved oxygen	≥4 (monthly mean) ≥3 (individual sample)	Near saturation
Fluoride*	Narrative	Narrative
Hardness	Narrative	Narrative
Iron (filterable)	0.3	Virtually absent
Lead*	0.05	Absent
Manganese (filterable)*	0.05	Absent
Nitrates plus nitrites*	10 (as N)	Virtually absent
pH (range)	6-8.5	Narrative
Phosphorus*	Narrative	Narrative
Selenium*	0.01	Absent
Silver*	0.05	Absent
Sulfate*	250	<50
Total dissolved solids (filterable residue)*	500	<200
Uranyl ion*	5	Absent
Zinc*	5	Virtually absent
Organic chemicals		
Carbon chloroform extract (CCE)*	0.15	<0.04
Cyanide*	0.20	Absent
Methylene blue active substances*	0.5	Virtually absent
Oil and grease*	Virtually absent	Absent
Pesticides		
Aldrin*	0.017	Absent
Chlordane*	0.003	Absent
DDT*	0.042	Absent
Dieldrin*	0.017	Absent
Endrin*	0.001	Absent
Heptachlor*	0.018	Absent
Heptachlor epoxide*	0.018	Absent
Lindane*	0.056	Absent
Methoxychlor*	0.035	Absent
Organic phosphates plus carbamates*	0.1‡	Absent
Toxaphene*	0.005	Absent
Herbicides		
2,4-D, 2,4,5-T, plus 2,4,5-TP*	0.1	Absent
Phenols*	0.001	Absent
	(Particle count per liter)	(Particle count per liter)
Radioactivity		
Gross beta*	1000	<100
Radium 226*	3	<1
Strontium 90*	10	<2

Air pollution series of technical publications

Environmental Protection Agency
Air Pollution Control Office
Office of Technical Information and Publications
Air Pollution Technical Information Center
P. O. Box 12055
Research Triangle Park, North Carolina 27709

ORDERING INSTRUCTIONS: If a paper copy is desired and a Government Printing Office price is given, the item MUST be ordered from the Government Printing Office using the GPO number listed. If no GPO price is given, a paper copy can be ordered from the National Technical Information Service, by use of the NTIS number. NTIS CANNOT sell paper copies if they are available from GPO. If a microfiche copy is desired, the order MUST be sent to NTIS. Payment must accompany all orders.

GPO ADDRESS	NTIS ADDRESS	NTIS PRICES
		PAPER COPY:
Superintendent of Documents	National Technical Information Service	1-300 pages = $3.00
Government Printing Office		301-600 pages = $6.00
Washington, D. C. 20402	U. S. Department of Commerce	601-900 pages = $9.00
	5285 Port Royal Road	Microfiche: $0.95
	Springfield, Virginia 22151	

Title	GPO number	GPO price	NTIS number
Air pollution aspects of the iron and steel industry	FS2.300:AP-1	—	PB 168867
Atmospheric emissions from fuel oil combustion—an inventory guide	FS2.300:AP-2	—	PB 168874
A pilot study of air pollution in Jacksonville, Fla.	FS2.300:AP-3	—	PB 168888
Air pollution and the Kraft Pulping Industry—an annotated bibliography	FS2.300:AP-4	—	PB 170744
Dynamic irradiation chamber tests of automotive exhaust	FS2.300:AP-5	—	PB 168877
Methods of measuring and monitoring atmospheric sulfur dioxide	FS2.300:AP-6	—	PB 168865
The November-December 1962 air pollution episode in the eastern United States	FS2.300:AP-7	—	PB 168878
A study in air pollution in the interstate region of Lewiston, Idaho, and Clarkston, Washington	FS2.300:AP-8	—	PB 168866
Air pollution in the coffee roasting industry	FS2.300:AP-9	$0.20	PB 168876
Community perception of air quality—an opinion survey in Clarkston, Washington	FS2.300:AP-10	—	PB 168875
Selected methods for the measurement of air pollutants	FS2.300:AP-11	—	PB 169677

Air pollution series of technical publications—cont'd

Title	GPO number	GPO price	NTIS number
Survey of lead in the atmosphere of three urban communities	FS2.300:AP-12	—	PB 170739
Atmospheric emissions from sulfuric acid manufacturing processes	FS2.300:AP-13	$0.60	PB 190235
Reactivity of organic substances in atmospheric photooxidation reactions	FS2.300:AP-14	—	PB 168879
Symposium environmental measurements—valid data and logical interpretation	FS2.300:AP-15	—	PB 168791
Potential dispersion of plumes from large power plants	FS2.300:AP-16	—	PB 168790
Atmospheric emissions from the manufacture of portland cement	FS2.300:AP-17	—	PB 190236
An air resource management plan for the Nashville metropolitan area	FS2.300:AP-18	—	PB 170740
The trend of suspended particulates in urban air—1957-1964	FS2.300:AP-19	—	PB 170475
Effects of the ratio of hydrocarbon to oxides of nitrogen in irradiated auto exhaust	FS2.300:AP-20	—	PB 190237
Continuous air monitoring program, Cincinnati 1962-1963	FS2.300:AP-21	—	PB 168863
Air particulates no. 1. Study number 22, Analytical Reference Service	FS2.300:AP-22	—	PB 170701
Continuous air monitoring program, Washington, D. C. 1962-1963	FS2.300:AP-23	—	PB 173987
Atmospheric emissions from coal combustion— an inventory guide	FS2.300:AP-24	—	PB 170851
Seminar on human biometeorology	FS2.300:AP-25	—	PB 190238
Workbook of atmospheric dispersion estimates	FS2.300:AP-26	—	PB 191482
Atmospheric emissions from nitric acid manufacturing processes	FS2.300:AP-27	—	PB 190239
Air pollution aspects of tepee burners	FS2.300:AP-28	—	PB 173986
Rapid survey technique for estimating community air pollution emissions	FS2.300:AP-29	—	PB 190240
Optical properties and visual effects of smokestack plumes	FS2.300:AP-30	—	PB 174705
Control and disposal of cotton ginning wastes	FS2.300:AP-31	$0.50	PB 174427
Selection and training of judges for sensory evaluation of the intensity and character of diesel exhaust odors	FS2.300:AP-32	—	PB 174707
Sources of polynuclear hydrocarbons in the atmosphere	FS2.300:AP-33	$0.30	PB 174706
Climatology of stagnating anticyclones east of the Rocky Mountains 1963-1965	FS2.300:AP-34	—	PB 174709

Air pollution series of technical publications—cont'd

Title	GPO number	GPO price	NTIS number
Emissions from coal-fired power plants	FS2.300:AP-35	$0.25	PB 174708
Health aspects of castor bean dust—review and bibliography	FS2.300:AP-36	—	PB 190241
Symposium on power systems for electric vehicles	FS2.300:AP-37	—	PB 177706
Pilot study of ultraviolet radiation in Los Angeles—October 1965	FS2.300:AP-38	$0.45	PB 182261
United States metropolitan mortality, 1959-1961	FS2.300:AP-39	—	PB 190242
Air pollution engineering manual	FS2.300:AP-40	?	PB 190243
Calculating future carbon monoxide emissions and concentrations from urban traffic data	FS2.300:AP-41	$0.50	PB 190244
Compilation of air pollutant emission factors	FS2.300:AP-42	$0.35	PB 190245
A compilation of selected air pollution emission control regulations and ordinances	FS2.300:AP-43	—	PB 190246
Handbook of air pollution	FS2.300:AP-44	$2.25	PB 190247
Thanksgiving 1966 air pollution episode in the eastern United States	FS2.300:AP-45	$0.30	PB 190248
Interim guide of good practice for incineration at federal facilities	FS2.300:AP-46	—	PB 190249
Guide to research in air pollution, 1969	FS2.300:AP-47	$1.50	PB 192220
Atmospheric emissions from thermal-process phosphoric acid manufacture	FS2.300:AP-48	$0.40	PB 190250
Air quality criteria for particulate matter	FS2.300:AP-49	$1.75	PB 190251
Air quality criteria for sulfur oxides	FS2.300:AP-50	$0.50	PB 190252
Control techniques for particulate air pollutants	FS2.300:AP-51	$1.75	PB 190253
Control techniques for sulfur oxide air pollutants	FS2.300:AP-52	$1.25	PB 190254
St. Louis dispersion study, Volume II—Analysis	FS2.300:AP-53	—	PB 190255
Atmospheric emissions from hydrochloric acid manufacturing processes	FS2.300:AP-54	$0.35	PB 190256
Tobacco, a sensitive monitor for photochemical air pollution	FS2.300:AP-55	$0.25	PB 190257
Air pollution translations: A bibliography with abstracts. Volume I	FS2.300:AP-56	$1.75	PB 190258
Atmospheric emissions from wet-process phosphoric acid manufacture	FS2.300:AP-57	$0.45	PB 192222
Air pollution aspects of brass and bronze smelting and refining industry	FS2.300:AP-58	$0.35	PB 190259
The climate of cities	FS2.300:AP-59	$0.55	PB 190260
Sensory evaluation of diesel exhaust odors	FS2.300:AP-60	$0.70	PB 192224
Characteristics of particulate patterns 1957-1966	FS2.300:AP-61	$0.50	PB 192223

Air pollution series of technical publications—cont'd

Title	GPO number	GPO price	NTIS number
Air quality criteria for carbon monoxide	FS2.300:AP-62	$1.50	PB 190261
Air quality criteria for photochemical oxidants	FS2.300:AP-63	$1.75	PB 190262
Air quality criteria for hydrocarbons	FS2.300:AP-64	$1.25	PB 190489
Control techniques for carbon monoxide emissions from stationary sources	FS2.300:AP-65	$0.70	PB 190263
Control techniques for Co, NO_x, and hydrocarbon emissions from mobile sources	FS2.300:AP-66	$1.25	PB 190264
Control techniques for nitrogen oxide emissions from stationary sources	FS2.300:AP-67	$1.00	PB 190265
Control techniques for hydrocarbon and organic solvent emissions from stationary sources	FS2.300:AP-68	$1.00	PB 190266
Air pollution translations: A bibliography with abstracts. Volume II	FS2.300:AP-69	$1.00	PB 196174
Highlights of selected air pollution research grants	FS2.300:AP-70	$0.70	PB 191273
Air pollution injury to vegetation	FS2.300:AP-71	$1.25	PB 193480
Nitrogen oxides: An annotated bibliography	FS2.300:AP-72	$2.75	PB 194429
Nationwide inventory of air pollutant emissions (1968)	FS2.300:AP-73	$0.30	PB 196304
Economic impact of air pollution controls on gray iron foundry industry	FS2.300:AP-74	$0.65	—
Hydrocarbons and air pollution: An annotated bibliography	FS2.300:AP-75	$5.00	PB 197165

NOTE: GPO prices listed are those in effect at time of writing but are subject to change.

192

J
Federal pollution control agencies

National Air Pollution Control Administration
U. S. Department of Health, Education, and Welfare
801 North Randolf Street
Arlington, Virginia 22203

Environmental Control Administration
Consumer Protection and Environmental Health
  Service
222 East Central Parkway
Cincinnati, Ohio 45202

Federal Water Pollution Control Administration—
  regional offices
Middle Atlantic
  318 Emmet Street
  Charlottesville, Virginia 22901
Southeast
  1421 Peachtree Street NE
  Atlanta, Georgia 30309
Ohio Basin
  4676 Columbia Parkway
  Cincinnati, Ohio 45226
Northeast
  John F. Kennedy Federal Building
  Boston, Massachusetts 02203
Great Lakes
  33 East Congress Parkway
  Chicago, Illinois 60605
South central
  1402 Elm Street
  Dallas, Texas 75202
Southwest
  100 McAllister Street
  San Francisco, California 94102
Northwest
  Pittock Block
  Portland, Oregon 97205
Missouri Basin
  911 Walnut Street
  Kansas City, Missouri 64106
Consumer Protection and Environmental Health
  Service—regional offices
Region 1
  John F. Kennedy Federal Building
  Boston, Massachusetts 02203
  (Connecticut, Maine, Massachusetts, New
    Hampshire, Rhode Island, Vermont)

Region 2
  26 Federal Plaza
  New York, New York 10007
  (Delaware, New Jersey, New York, Pennsylvania)
Region 3
  220 7th Street NE
  Charlottesville, Virginia 22901
  (District of Columbia, Kentucky, Maryland,
    North Carolina, Virginia, West Virginia,
    Puerto Rico, Virgin Islands)
Region 4
  50 7th Street NE, Room 404
  Atlanta, Georgia 30323
  (Alabama, Florida, Georgia, Mississippi, South
    Carolina, Tennessee)
Region 5
  433 West Van Buren Street
  New Post Office Building, Room 712
  Chicago, Illinois, 60607
  (Illinois, Indiana, Michigan, Ohio, Wisconsin)
Region 6
  601 East 12th Street
  Kansas City, Missouri 64106
  (Iowa, Kansas, Missouri, Minnesota, Nebraska,
    North Dakota, South Dakota)
Region 7
  1114 Commerce Street
  Dallas, Texas 75202
  (Arkansas, Louisiana, New Mexico, Oklahoma, Texas)
Region 8
  9017 Federal Office Building
  19th and Stout Street
  Denver, Colorado 80202
  (Colorado, Idaho, Montana, Utah, Wyoming)
Region 9
  Federal Office Building
  50 Fulton Street
  San Francisco, California 94102
  (Alaska, Arizona, California, Nevada, Hawaii,
    Oregon, Washington, Guam, American
    Samoa)